Contents

Ballistic Missile Proliferation in the Third World

INTRODUCTION

At a time when major reductions in NATO and Warsaw Pact (WP) strategic forces seem probable, a new threat to international security is emerging: an increasing number of Third World states are acquiring ballistic missile capabilities.[1] In March 1989 the Director of the United States Central Intelligence Agency, William Webster, warned that at least 15 Third World countries could be manufacturing their own ballistic missile systems by the turn of the century.[2] There is a growing concern amongst both policy-makers and academics that ballistic missile proliferation and improvements in Third World missile payload, range, reliability and accuracy have important and potentially adverse implications for the arms race in the regions, for crisis stability and for international security.

The dangers of ballistic missiles derive from their capabilities. First, ballistic missiles can attack distant targets with a rapidity that in many cases precludes warning. Second, ballistic missiles are very difficult to destroy once in flight. Third, ballistic missiles can carry a variety of destructive munitions, including chemical and nuclear ordnance. As a result, ballistic missile proliferation could make Third World crises more difficult to control and conflicts deadlier than ever.

The spread of ballistic missiles may have significant consequences for stability in many Third World regions. In a crisis situation, stability could be undermined as each state attempts to secure military objectives and prevent damage to its homeland by getting in the first blow. Fears of missile attacks against either military targets or cities could lead states to adopt pre-emptive strikes against an enemy's missile capabilities. The dangers, of course, may be even greater should missiles be deployed with chemical or nuclear warheads. It is significant that the areas of the most intense missile proliferation are also those of the greatest instability, namely the Middle East and South Asia. Ballistic missile proliferation could therefore aggravate already tense political and military situations.

It has, however, been argued with regard to the spread of nuclear weapons that proliferation processes can have stabilizing consequences and that 'more may be better'.[3] Indeed, it has been suggested that nuclear deterrence, which has kept the peace between the superpowers and their allies for over 40 years, might work in Third World settings as well. Following this line of thinking, caution and mutual restraint might be reinforced by nuclear proliferation and perhaps by the proliferation of other advanced military technologies.

Nevertheless, this study contends that ballistic missile proliferation is disturbing, to say the least. Deterrence has never made war between the superpowers impossible, only irrational.[4] Moreover, the concept of deterrence and the desirability of a stalemate imposed by mutually assured destruction have been contested within the American and Soviet strategic communities. Little is known about strategic thinking in many proliferating states, and it is quite possible that perceptions may vary greatly there. In any event, many Third World countries presently face a far less predictable strategic environment than did the superpowers during most of their years of rivalry. In the developing nations, far less is usually known about enemy intentions or capabilities. In addition, there are fewer resources and less time to invest in the kind of strategic analysis undertaken in the West. Furthermore, in the Third World, where conflicts may be more intense and willingness to suffer casualties may be greater, mutual deterrence could be sorely tested by the proliferation of high-speed delivery capabilities.

Another cause for concern is the relationship between the spread of missile technologies and non-conventional ordnance. There seems to be a correlation between states intent upon acquiring ballistic missiles and those seeking to develop arsenals of chemical and, in some cases, nuclear weapons. A number of developing states have already used missiles and chemical weapons, and it is not unreasonable to assume that delivery vehicles and non-conventional ordnance could be employed in a future conflict. It is also possible that ballistic missile proliferation could heighten Third World interest in acquiring chemical and nuclear weapons. It might seem highly advantageous to endow ballistic missiles with non-conventional capabilities.

The spread of ballistic missiles may also have important consequences for the industrialized world. The merging of long-range ballistic systems with highly destructive ordnance only serves to underline that the consequences of ballistic missile proliferation cannot be confined to particular regions. In any event, the reduction of East–West tensions and the decline of Soviet global influence indicates that Third World defence policies and security concerns may play a much bigger role in international security affairs in the years ahead. Third World attempts to acquire sophisticated weapons such as ballistic missile systems are a clear manifestation of the developing nations' desire to increase their military and political influence and, indeed, reflects the evolving multipolarity of the international order.

Ballistic missile proliferation also poses a number of significant problems for arms control. It is possible to agree that the spread of ballistic missiles is a subject of import while arguing that it is neither necessary nor desirable to treat the topic as a distinct one. As noted, the proliferation of ballistic missiles is inextricably linked with the spread of nuclear, chemical and conventional arms. Indeed, from the perspective of foreign policy in general, and arms control in particular, the issue can be treated as subordinate to such other concerns as nuclear proliferation and conventional arms transfers.

Nevertheless, this study contends that it is necessary here to focus on delivery systems rather than ordnance and, in particular, to differentiate between ballistic missiles and other delivery systems. The processes by which states procure or supply weapons and the means by which these processes can best be controlled may be system specific in many important respects. Moreover, the motivations behind ballistic missile proliferation and the objectives to which these missiles might be put need not be strongly related to the issue of nuclear proliferation. Indeed, it is important to view the problem not only from a Western standpoint but from a Third World perspective – a perspective from which the values and concerns of Western analysts do not always predominate. Thus, while nuclear and chemical proliferation cannot be ignored in a study of ballistic missile proliferation, the latter will be treated as a discrete subject in this Paper. The spread of cruise missiles, another problem in the Third World, will therefore not be included in this study.

The need to analyse the processes underlying the spread of ballistic systems in Third World states thus emanates from a number of sources, including concerns over regional stability as well as about the problem of maintaining international security in an increasingly multipolar international system. The implications of the spread of ballistic missiles extend far beyond the expansion of Third World firepower and capabilities. Indeed, it is one reflection of the intermeshing of North and South security concerns. It is for these reasons that ballistic missile proliferation deserves the attention of arms controllers in both the United States and Europe.

The issues to be addressed in this study concern both the motivations behind the Third World's acquisition of ballistic missiles and the supply processes which enable proliferation to occur. The effect of the spread of missile systems on the strategic policies of some of the countries most involved in missile development and/or deployment will also be discussed. So, too, will the broader regional and international implications of expanding capabilities to threaten enemy targets with a variety of lethal ordnance. Finally, an analysis of various arms-control options will be presented, with the goal of identifying the best way to proceed. By drawing together the political and strategic issues associated with ballistic missile proliferation and its control, this study will seek to advance the evolving debate on this new threat to international security.

This Paper focuses on common motivations and processes underlying ballistic missile proliferation rather than specific missile deployment numbers. Proliferation remains an extremely sensitive issue for Third World governments, and it is therefore difficult to draw up a reliable data base of systems deployed and programmes under development. Nevertheless, more information on Third World systems is steadily becoming available in the unclassified literature, and it is possible to identify the major outlines of many Third World projects.

Chapter I will examine the causes of Third World ballistic missile proliferation. Chapter II will analyse the regional and international implications of this process, while Chapter III will discuss the problems associated with ballistic missile arms control and identify the preferred policy options for the West in the 1990s.

I. THE SOURCES OF BALLISTIC MISSILE PROLIFERATION IN THE THIRD WORLD

Introduction

The global spread of ballistic missiles must be viewed against the background of the increasing military and political autonomy of smaller states and their significant progress in developing scientific expertise and military–industrial capabilities.

An examination of the rationales of Third World countries regarding their acquisition of ballistic missiles reveals no single predominating motivation. Rather, these countries seem to have a variety of incentives and objectives, ranging from enhancing national prestige to military employment. The processes whereby developing states acquire ballistic missiles are also complex. Most Third World missiles have been bought from foreign suppliers, but an increasing number of countries are attempting to develop their own systems. In order to facilitate such projects, many Third World missile developers still require external assistance. The importation of important missile components and the employment of foreign expertise have been crucial features of the majority of Third World missile programmes.

At the beginning of the missile proliferation era in the 1960s, the supply of missiles was restricted to a flow between the superpowers and selected clients. In the 1970s and 1980s, however, the widening market for ballistic missiles and missile-related technologies led to an increase in the number and types of suppliers. These now include the industrialized states of Europe, the so-called second-tier suppliers in the Third World, and private consortiums based in Europe.

Thus the spread of ballistic missiles has been transformed into an extremely complex process, driven by a wide variety of military, political and economic goals, and facilitated by a multifarious transfer market.

The scope of proliferation

The spread of ballistic missile systems is a truly global Third World phenomenon. Developing states in the Middle East, South Asia, East Asia, Latin America and Southern Africa either possess or have demonstrated the intent to acquire ballistic missiles with the object of deploying and, in some cases, marketing these systems. The most concentrated area of proliferation is centred in the region stretching from the Levant to the Maghreb, where Iran, Iraq, Saudi Arabia, Kuwait, the Yemen Arab Republic, the People's Democratic Republic of Yemen,[1] Syria, Algeria, Israel, Egypt and Libya all possess ballistic missiles. In Southern Asia there is a narrower but equally intense build up of missile capabilities, as reflected in the expanding arsenals of India and Pakistan. Afghanistan, too, has acquired ballistic systems. In East Asia, North Korea, South Korea and Taiwan have ballistic missile systems, while Indonesia has already tested a number of sounding rockets. In Central America, Cuba possesses missile batteries while to the south, Brazil and Argentina have engaged in sophisticated ballistic missile development projects. Across the Atlantic, South Africa has tested a ballistic missile.[2]

Twenty-two countries are thus identified as Third World missile proliferators in this study – that is, states which already deploy or are attempting to develop ballistic missiles. They are: Argentina, Brazil, India, Saudi Arabia, Israel, Iraq, Taiwan, Pakistan, Indonesia, South Africa, Egypt, Iran, Cuba, Syria, Libya, North Yemen, South Yemen, North Korea, South Korea, Afghanistan, Algeria and Kuwait. Three general missile types, broadly distinguishable according to their ranges,[3] are involved in this proliferation process. (See Tables on pp. 29–31).

THIRD WORLD LONG-RANGE MISSILES (TWLRM) > 900 KM

During the past decade a number of Third World states began manufacturing rocket launchers which are capable of delivering warheads over great distances, and in some cases, thousands of kilometres. Some Third World countries have invested in sounding rockets and powerful space launchers which can be modified for long-range surface-to-surface delivery. Once deployed, some of the longer-range versions may provide developing nations with intercontinental delivery capabilities. TWLRMs remain few in number, being technically difficult to produce. They often require, for example, complex guidance and control systems, as well as heat shielding for re-entry purposes. TWLRM systems are rarely available on the open market. Countries which have successfully tested TWLRMs are India, Israel and Iraq. Saudi Arabia deploys the Chinese-built CSS-2.

THIRD WORLD MEDIUM-RANGE MISSILES (TWMRM) 200–900 KM

Much of Third World missile procurement involves systems of much shorter ranges but which can be employed as strategic weapons in regional contexts, as they have sufficient range to attack enemy cities and military installations beyond the battlefield. Many indigenously produced missiles and purchased systems which have been modified domestically are also found in this category. Countries which already deploy TWMRMs are Afghanistan, Israel, Iraq, Iran, Egypt, Syria, Libya, South Yemen, Pakistan, North Korea and South Korea.

THIRD WORLD SHORT-RANGE MISSILES (TWSRM) < 200 KM

In the Third World, short-range ballistic missiles serve battlefield functions. Many of these systems are among the oldest and most common in the Third World, since they have been the easiest to buy and the least demanding to develop. As these weapons do not necessarily require sophisticated guidance or propulsion technologies, many countries also manufacture their own domestic models. A number of developing countries manufacture their own rocket systems, whose technologies may be used to develop ballistic missiles.[4] Countries which deploy TWSRMs are Libya, North Yemen, Syria, Iraq, Israel, Egypt, Iran, Kuwait, South Yemen, Taiwan, Cuba, North Korea, Pakistan, Algeria and South Korea.

The significance of ballistic missile proliferation derives not only from the increasing military reach which they provide Third World states, but also from the types of ordnance they may deliver. Ballistic missiles can deliver non-conventional ordnance as well as high explosive warheads. Indeed, an analysis of the scope of ballistic missile proliferation and that of the spread of non-conventional ordnance reveals that a correlation exists between those states seeking to acquire ballistic missiles and those pursuing chemical, biological and nuclear programmes.

Of the 22 countries under study, three (Israel, India and Pakistan) are believed to have nuclear weapons in stock or ready for easy assembly, while four others (Argentina, Brazil, South Africa and Iraq) have significant nuclear programmes. More specifically, there is a correlation amongst these countries between the development of nuclear programmes and the development of TWLRMs – though of course this is not to state that the two are causally connected. There is a further correlation between missile and chemical/biological proliferation. Ten of these countries – Egypt, Iran, Iraq, Israel, Libya, North Korea, South Korea, South Africa, Syria and Taiwan – are believed to have stockpiles of chemical weapons, while four – Iran, Iraq, North Korea and Syria – are alleged to hold stockpiles of biological weapons.[5] Estimates on non-conventional ordnance may, however, vary.

An interest in the development and/or procurement of ballistic systems does not necessarily imply an interest in employing missiles for non-conventional missions or a capacity to manufacture nuclear, chemical and biological warheads for them. However, it is the potential to do so that raises concerns about the spread of ballistic missiles in the Third World. It is therefore of importance to analyse the motivations and processes underlying the proliferation of ballistic systems.

Third World rationales for ballistic missile acquisition

One way of determining why nations acquire missile systems is to analyse relevant statements emanating from the proliferating states themselves. This may help to reveal the specific perceptions of these countries regarding their need for ballistic missiles, or at least highlight the impression they may wish to create as to why they are obtaining surface-to-surface systems.

In analysing declaratory statements it should be noted that many of these pronouncements may reflect rationalizations and legitimizations rather than true national motivations; some statements may even be intentionally misleading. However, legitimizations and motivations are often not readily separable into distinct categories. Moreover, all forms of declaratory statements may be incorporated into the calculations of regional competitors and are therefore of interest in themselves.

An analysis of pronouncements by Third World states concerning their motivations for acquiring ballistic missiles reveals no single predominant theme. Rather, there is a spectrum of rationales stretching from references to the prestige associated with the possession of ballistic systems to an emphasis on their military utility. The attractions of status and prestige are elements to be found in the declaratory pronouncements of a wide variety of missile proliferating states, but they are clearly most evident in countries where regional and international prestige are regarded as major foreign-policy goals. This also appears to be true where there is a strong incentive to market missiles but a necessity to avoid explicit reference to such a fact: no state openly maintains that its ballistic missile development programme is primarily directed at exports. A focus on status is more evident amongst states domestically developing missiles rather than countries simply procuring them, the latter of which cannot take the credit for the technological expertise involved in building the systems.

In many cases, the rationale for missile acquisition is more than prestige. Some Third World states maintain that they are acquiring missiles for general balance of power reasons, or for deterrence against specific threats (either emanating from enemy power in general or enemy missiles in particular) or in order to employ them for war fighting purposes. In declaratory policy, there is a definite preference for emphasizing the conventional as opposed to the nuclear and/or chemical potential of ballistic missiles, although oblique references to ballistic missiles' non-conventional delivery potential are sometimes made.

PRESTIGE

In the Third World, ballistic missiles and satellite launchers have come to be regarded as important symbols of technological achievement and military prowess, aside from any specific military task to which they could be assigned. The ability to master the complexities of guidance technology, rocket fuels and metallurgy reflects a state's technological sophistication and competence and is therefore clearly attractive to many developing countries. For many Third World countries, missiles serve as 'technological demonstrators' or symbols of domestic technical competence.

Status rationales accompanying the acquisition of ballistic systems have been especially pronounced in the cases of India, Brazil and Argentina. In the 1970s these states invested heavily in rocket systems, ostensibly for scientific and space-related reasons. Domestically and internationally, they portrayed their research as an indication of their scientific sophistication and advancement. However, by the 1980s, space research was increasingly accompanied by military research activities, with both depicted as indications of advanced capabilities, technological self-sufficiency and military independence. Thus, in India, following the launch of the 2,500-km *Agni* in 1989, one official stressed that the '*Agni* is a manifestation of India's self reliance and a reiteration of faith in our capabilities . . . China can reach our territory at any time. Can any self-respecting country accept that?'[6]

These states believe that the prestige and status associated with such projects as rocket research will help them to achieve broad political objectives, particularly an expanded regional role and greater attentiveness to their concerns on the part of the more powerful nations.

This tendency is also found in South Africa, where foreign policy has been geared towards staving off isolation and aligning itself with the industrialized West. Rocket research has been used to emphasize 'the high level of technological advancement in South Africa, a level that is remarkable for a small nation that is rooted in a predominantly Third World environment'.[7] By stressing its advanced technological capabilities, Pretoria wishes to enhance its respectability in Western eyes.

Prestige and military capability are often intermeshed. Indeed, even in a country such as Iran, where the expressed rationales for ballistic systems acquisition are strongly military in nature, policy-makers do not shy away from the technical and military prestige associated with missile development. For example, an Iranian government official speaking on Teheran's domestic radio service proudly quoted a Japanese magazine that referred to Iran as Asia's second missile power.[8] More specifically, the technical linkage between space and missile programmes means that a country can emphasize the prestige associated with its space research and, at the same time, hint at its military potential and deterrent capability. In Israel – a country which does not admit to deploying TWLRMs or TWMRMs, but whose motivations for missile and space research are predominantly military – prestige and status are also stressed. Following Israel's satellite launch in 1988, Prime Minister Itzhak Shamir stated that it 'places Israel among the world's technological elite'.[9]

Clearly, where there is little military rationale for missile research and deployment, prestige and status are very important motivations. Countries such as Brazil and South Africa are the clearest examples of this, as neither have obvious targets against which to direct missiles. In the Brazilian case, however, export possibilities may also be a major factor. What is obvious, however, is that while prestige and status are not the main objectives for many proliferating countries, there appears to be a widespread view among these states that missile research can serve political goals.

DETERRENCE AND WAR FIGHTING

Third World states recognize that missiles can serve military objectives. However, within the declaratory policies of Third World states there are varying degrees of explicitness with regard to the nature and sources of the threats being deterred and the manner in which missiles could be employed.

The development of missiles for the purpose of deterring enemy use of missiles, as well as for war fighting, is most clearly illustrated in the case of Iran. In March 1988, during the 'war of the cities', Iranian Prime Minister Hashemi Rafsanjani stressed that Iran's main means of dealing with the Iraqi missile threat was to bolster its own missile industries. Iranian policy was to ensure that 'the very thought of an attack with missiles will be eliminated from our neighbour's minds'.[10] Elsewhere, the Prime Minister stated that 'if we pay more attention to our missile industry, we can get the better of Iraqi missiles'. He stressed that Iran did not wish to pursue such a course, but that it needed to dissuade the Iraqi government from attacking Iranian cities.[11]

In Pakistani declaratory statements, reference has been made to the need to deploy missiles for the purpose of deterring enemy missiles in particular, and aerial delivery capabilities in general. When the Minister of State for Defence, Ghulam Sarwar Cheema, was asked why Pakistan was developing ballistic missiles, he replied that the country had 'to have an antidote for what our enemy [India] next door has'.[12] In the case of Saudi Arabia, the government has explained that it needs missiles for defence against Iranian attacks.[13] The Saudis implied that their missiles would be launched in response to a wide variety of Iranian attacks, not just Iranian missile assaults.[14] Since the end of the Iran–Iraq War, the Iraqis have frequently claimed that the role of their missiles is to deter any future Israeli missile or air attack. The most forthright Iraqi statement was made by President Saddam Hussein, who warned: 'I swear to God, we will let our fire scorch half of Israel if it tries to wage anything against us!'[15] It is generally accepted that the strength of Israel's air defence system makes ballistic missiles the only viable Iraqi option for attacking the Jewish state.

The clearest example of a missile arms race dynamic is that of Iran and Iraq, where the appearance of enemy missiles reinforced pressures to deploy missiles in response. Declaratory statements indicate that this may also be true of the Pakistan–India case, at least from Karachi's perspective. In other areas, the motivations behind missile deployment appear to be more diffuse; missiles serve as a response to a wide range of enemy capabilities and actions.

Some countries prefer to be less explicit about which states they are interested in influencing with their missile forces. This may stem from the need to avoid aggravating regional tensions and, in some cases, due to a lack of strategically worthwhile targets. India, which prefers not to antagonize the People's Republic of China (PRC) and Pakistan unnecessarily, has often referred to its missile force not solely in terms of deterring specific countries, but on a more general level. For example, Air Commodore Jasjit Singh has pointed out that the 'time has come for the Indian air force to think of integrating its deep strike aircraft with a possible ballistic missile force to constitute a strategic deterrent capability for national defence.'[16] This statement is broad enough to cover all threats to India's national security.

Many observers believe that Syria has compelling military reasons for acquiring ballistic missiles. Yet, in its declaratory policy, Damascus does not overly stress its missile capabilities nor does it specify the military objectives which these systems are supposed to further. This may reflect the recognition that too strong an emphasis on the missile's role in war planning could be counterproductive: it could reinforce and legitimize Israeli military responses (such as pre-emptive strikes against missile launch sites).

Israel, in turn, is quite explicit that its deterrent is directed against its Arab enemies, but is less open about the make up of its deterrent force. Although a number of non-Israeli sources maintain that Israel possesses medium-range missiles and is developing long-range systems, Israel prefers to frame its deterrent posture in a way that neither confirms nor denies the existence of missile or non-conventional capabilities. This does not prevent Israel from threatening massive retaliation should enemy missiles be launched at its territory. Thus, former Minister of Defence Itzhak Rabin has been unequivocal in saying that 'Israel not only has the deterrent power to face down the missiles held by Syria, Iraq and now also Saudi Arabia . . . we also have the ability to attack their populated areas to a degree that outstrips theirs many times over'.[17] Clearly, Israel's long-range missile programmes are influenced by growing Arab missile capabilities, although air power remains Israel's most effective long-range delivery system.[18]

Most states avoid emphasizing any linkage of their missile and nuclear or chemical programmes. Indian officials and commentators often stress that New Delhi's missile programme is not geared to nuclear delivery.[19] Following its acquisition of the DF-3 missiles from the PRC, Saudi Arabia signed the Non-Proliferation Treaty (NPT) in order to reassure the outside world of its intentions in this area. Riyadh has also reaffirmed its opposition to the use of chemical weapons.[20] Prior to their recent announcement that they would suspend their missile programme, military sources in Argentina stated that their missiles were not intended to deliver nuclear ordnance.[21] Israel has hinted at its willingness to use non-conventional ordnance, possibly delivered by surface-to-surface missiles, if its enemies use similar weapons against it.[22] By the same token, Iraq has not sought to dampen speculation that it would use chemical weapons in the event of an Israeli attack.[23]

A preference for discretion

In analysing declaratory pronouncements, it is important to recognize that most missile proliferators prefer to say nothing, or as little as possible, about their missile projects. This may indicate that rationales have not been carefully thought through, or that states wish to avoid giving further impetus to regional arms races. It is also likely, however, that proliferators are aware of international sensibilities and the vulnerability of domestic programmes to external interference.

In this sense, ballistic missile proliferation takes place in a manner akin to the proliferation of non-conventional weaponry. Similarly to nuclear and chemical weapons, missiles have been singled out as dangerous to regional and international security, and therefore items to be controlled. Acquisition must thus be carried out as unobtrusively as possible. Furthermore, in one country, Israel, the 'nuclear bomb in the basement' strategy is now being mirrored by a 'missile in the basement' approach: the benefits for deterrence and status are not compromised by uncertainties about the weapons' existence, while the arms race and political costs of declaring the capability are at least partially avoided.

Rather than being discreet, Iraq sometimes trumpets the existence of its missile programme. This must be viewed in the context that Baghdad has over the past few years not only ignored international norms on the employment of chemical weaponry, but appears to believe that calculated disregard of certain international standards strengthens its regional position.

Processes of acquisition

There are two methods by which a state may acquire ballistic missiles: domestic production and procurement. The capabilities of domestically produced missiles vary considerably from country to country, with the major independent variables being the general sophistication of local military industry and the degree of domestic technological competence. Importation of complete missile systems, relevant technologies and expertise has in the past served to counter the constraints imposed by limited domestic development capabilities. In addition, the importation of space technologies has sometimes helped Third World states to develop TWLRMs.

In order to deploy ballistic missiles, states may either purchase complete systems or develop their own weapons. The simplest method of acquiring ballistic missiles is to purchase them 'off the shelf' from more technologically advanced nations. Indeed, of the 22 countries examined in this study, 16 possess ballistic missiles that have been bought in a complete form from other states. These states are: Afghanistan, Algeria, Cuba, Egypt, Iran, Iraq, Israel, Kuwait, Libya, North Korea, South Korea, Saudi Arabia, Syria, North Yemen, South Yemen and Taiwan.[24] Eight of these – Afghanistan, Cuba, Saudi Arabia, Algeria, Syria, Kuwait, North Yemen and South Yemen – rely solely on imported systems. The missile proliferation process has undoubtedly been facilitated since the 1960s by the availability of complete missile systems to developing states. 'Off the shelf' weaponry accounts for the majority of missile systems currently deployed in Third World arsenals.

However, only a small number of missile types have been acquired by the Third World: various models of *FROGs*, *Scuds*, SS-21s, DF-3s, *Honest Johns* and *Lances*. The number of primary suppliers (as opposed to those who may retransfer missiles) has been smaller still: the Soviet Union, the United States and the PRC. Moreover, while domestic development involves more technological and economic challenges, over-the-counter purchases involve political problems that have increased as the issue of ballistic missile proliferation has gained greater international prominence. Thus, states wishing to procure 'off the shelf' missiles have had to accept limitations on range, payload, accuracy and quantity. Of the six missile types noted above, only the DF-3 can be described as a TWLRM, and in this case, the Saudis may have been provided with a variant of a missile that has limited range.[25] Furthermore, only the SS-21 is regarded as being relatively accurate and only the DF-3 and *Scud* have payloads over 1,000 kg.

Moreover, it is becoming increasingly difficult to buy missile systems on the international market at all. Suppliers have never been totally unconcerned about the transfer of sophisticated technology to the developing world, and together with evolving multilateral and bilateral pressures on missile transfers, Third World attempts to acquire 'off the shelf' missiles are often rejected. In the mid-1980s a Syrian request to buy Soviet SS-23s – a transfer agreement which would have subsequently breached the 1987 Intermediate Nuclear Forces (INF) accord – was rejected by Moscow.[26] Syria's desire to acquire Chinese-built M-9s continues to be frustrated.[27] Libya, too, has failed to secure Brazilian MB/EE missile technology and Soviet-built SS-21s and the 900-km SS-12.[28] Israel's decision in the early 1970s to purchase *Lance* missiles may have been taken only after the United States refused to provide the more capable *Pershing* 1A systems.[29] Saudi Arabia was denied *Lance* missiles by the US in 1985.[30] Iraq, in turn, may have made an unsuccessful attempt to arm itself with Soviet SS-12s.[31]

Procurement problems may encourage Third World states to develop their own missiles. Yet this option poses serious challenges of its own. The demands of manufacturing ballistic systems require the producer to master the complexities of rocket propulsion, guidance and control mechanisms, as well as (in longer-range systems) re-entry vehicle technology and, in the area of non-conventional ordnance, the adaptation of warhead load to the specific constraints of ballistic missile delivery.

Building key components and integrating them into functioning systems demands a level of industrial and technical expertise in rocketry, aeronautics, computer engineering, and microelectronics that is often far beyond that of developing states. Naturally, as demands for range, payload and accuracy increase, so do the requirements for technical proficiency. The large financial and technical investment involved in such an enterprise may serve as an important deterrent to domestic missile manufacture. Demands for resources to support ballistic weapon projects have to compete with demands for resources for other military and civilian projects. Finally, the attempt by many Western nations to curtail the spread of ballistic missile technologies makes the acquisition of key components, such as gyroscopes, accelerometers, ablative heat shields, thrust vector controls, solid rocket fuel, and composite technologies needed for rocket nozzles, a difficult enterprise, (though probably not as problematic as acquiring complete 'off the shelf' systems). Most of these components can be acquired from only a very small number of states – Italy, Japan, the UK, Canada, France, the US, the Federal Republic of Germany (FRG), Israel, the USSR and the PRC.[32]

Since the technological, economic and political complexities involved in developing missile systems are enormous, Third World programmes are never entirely indigenous. In many cases, developing countries upgrade systems which were originally obtained on the international arms market. Other Third World programmes rely on the importation of some missile subsystems, such as gyroscopes and rockets, as well as expertise in ballistic missile production.

The easiest method of developing a domestic system is to upgrade or reverse engineer a foreign-produced system already in one's possession. This allows the developing country to avoid manufacturing many of the complex components noted above and enables it to copy, and in some instances to improve, range and payload characteristics. As noted earlier, the missile that has been most extensively modified is the *Scud* B. This is hardly surprising considering that it is found in more Third World arsenals than any other ballistic missile. One country which has invested much effort in *Scud* modification has been North Korea.[33] Extensive upgrading of the *Scud* has also taken place in Iraq, where on 4 August 1987, Baghdad announced the test of the 600-km *Al-Husayn* missile and, on 29 February 1988, reported the successful launch of the 900-km *Al-Abbas* system.[34] Egypt has also improved its *Scuds*, while Iranian missile programmes have benefited greatly from *Scud* missile technology.

Other externally acquired systems have also been used as models for domestic weapons. The Israeli *Jericho* 1 missile is thought to have been developed jointly with France during the 1960s. It consists of Israeli improvements on the Avions Marcel Dassault MD-660 and MD-620.[35] The Taiwanese *Ching Feng* system is thought to be a copy of the American *Lance* missile.[36] India's SLV-3 solid propellant rocket bears a close resemblance to the American *Scout* developed in the 1950s.[37] In 1985, South Korea test-fired a ballistic missile based on a modified version of the US *Nike Hercules* surface-to-air missile.[38]

The level of foreign assistance in Third World missile development projects remains difficult to gauge due to the sensitivity of the subject. Third World missile developers often prefer to play down or reject claims that their programmes are benefiting from external assistance, thereby emphasizing their independent status as well as avoiding efforts to stop the foreign aid upon which their programmes are dependent. The requirement for technological assistance in the areas of electronics, machine tools and aerospace expertise becomes even greater in the case of longer-range and more capable weapon systems.

The most technologically advanced ballistic missile programme is in Israel which, in addition to the *Jericho* system, has built the highly accurate *Gabriel* MK-1V anti-ship weapon and the *Have Nap* ASM system. But even Israel required external technological and financial assistance to improve its *Jericho* 1 missile. Non-Israeli sources claim that the *Jericho* employs an inertial guidance system based on American technology.[39] Some have speculated that Israel's attempt to acquire a supercomputer from the US is at least partly motivated by a desire to modify its missile system.[40]

The need for expertise and manpower is even greater for states lacking Israel's technological prowess. For example, as one analyst pointed out, while Pakistan may maintain that its *Hatf* 1 and 2 are domestically produced systems, 'any indigenous [Pakistani] missile would have to consist almost entirely of foreign components assembled under foreign supervision'.[41] Iraq's *Condor* 2 programme was at least initially dependent on Argentinian, Egyptian and private West German co-operation. North Korea is said to have assisted Cairo's *Scud* upgrade programme,[42] while Libya's various missile programmes have been highly dependent on private European consortiums.[43] The PRC has recently entered into an agreement with Egypt to help Cairo with its missile programme.[44] South Africa's missile project is reportedly very dependent on Israeli assistance.[45]

Thus, states encumbered with technological constraints and facing political obstacles may still develop missile systems, provided that they are prepared to co-operate with foreign states or companies. Indeed, since the 1970s, while no domestically produced ballistic missile system can be said to be totally indigenous,[46] few Third World countries have been forced to abandon totally their missile research due to technical or political constraints. Indeed, although certain Third World missile projects – for example, in India, Egypt, Israel and Taiwan – have previously been cancelled, missile research leading to production and deployment has continued in all these states.[47] The ability to produce missiles is clearly well within the grasp of many developing countries, and thus problems faced in procuring 'off the shelf' systems will not necessarily stop the proliferation of Third World medium- and long-range missile capabilities.

Domestic production of satellite launchers and the development of long-range missiles

One significant feature of Third World ballistic missile programmes has been the development of long-range systems under the mantle of space launch programmes. Whenever a state acquires the ability to fire a payload into orbit and exercise control over its complete flight, it is well on the way to developing a militarily useful system.

The existence of a space programme or interest in developing such a programme is not necessarily an indication of interest in a military programme. [48] A wide range of benefits ranging from scientific data, a share in the space launcher market, and military intelligence may also be secured by pursuing a space programme. Nevertheless, there is a strong correlation in the Third World between an interest in space launchers and the desire to acquire long-range ballistic missiles. In Brazil, India and Israel, large investments in space programmes have been matched by huge investments in long-range missile systems. More recently in Iraq, space launcher research is accompanying long-range missile development programmes.

As many of the components for a space launcher, such as rocketry, propellant technologies, computers and launch facilities, are dual-capable, Third World countries can, if they so desire, disguise their military intentions under the mask of a civilian project. Dual-use technologies enable both the proliferator and supplier to legitimize technology transfer as part of a civilian activity.[49] Indeed, this approach may be necessary in order to circumvent efforts to prevent the spread of ballistic missiles.

Although Third World states may reject assertions that their space programmes are being pursued for military ends, they cannot (and do not) deny that civilian space launchers can be transformed into surface-to-surface ballistic missiles (if liquid-fuelled, albeit with dismal readiness alert and preparation characteristics).[50] As the Director of the Brazilian Aerospace Technical Centre, General Hugo Piva, stated, 'The development of [satellite launch rockets] will make us familiar with those techniques that are also used to manufacture weapons'.[51]

The relationship between space launchers and long-range missiles is obvious in a number of cases. The first stage of the Indian long-range *Agni* missile is the solid propellant of the Indian SLV-3 satellite launch vehicle.[52] In Brazil, there are striking similarities between the payloads of the *Sonda* space launcher series and Brazil's battlefield rocket series.[53] It is also probable that Brazil's MB/EE-1000 and the Avibras SS-1000 programmes are based on the *Sonda* 4 space launcher. In both the Indian and Brazilian cases, space research preceded the indigenous missile programmes by more than a decade.

On the other hand, Israel's military missile programme dates back to the 1960s, while the Israel Space Agency was only formed in 1983. One of the Agency's main tasks has been to adapt the *Jericho* 2 launcher to place satellites in orbit. Indeed, Israel's space programme followed on and derived benefit from existent missile programmes. Similarly, although Baghdad has been upgrading its *Scud* imports since the early 1980s, it was only in 1989 that it claimed to have fired a launcher capable of lifting a satellite into orbit.[54]

Two other countries in which space and missile research appear to be inextricably linked are Pakistan and Argentina. It has been claimed that 'the spur of [Pakistan's] space programme is the prospect of its long-term military implications'.[55] In Argentina, the Instituto de Investigaciones Aeronauticas y Espaciales undertakes space research and has also conducted research into the military employment of rockets. The *Condor* 1, launched in the early 1980s, was tested both as a sounding rocket (a rocket used for atmospheric test purposes) and as a surface-to-surface missile.[56]

While one should not underestimate the difficulties in converting a sounding rocket or a space launcher into a long-range missile, a number of technologically advanced Third World states seem to be confident that these technical hurdles can be overcome, albeit with the help of foreign expertise and importation of key technologies. However, it remains difficult to develop guidance and control systems, acquire propellants, and manufacture re-entry vehicles. In addition, the massive costs involved in, and growing international reservations about such programmes make it unlikely that investment in space launchers will be a popular way of developing long-range military systems in many Third World countries.

The supply of ballistic missiles

Since the 1950s, the arms transfer market has become increasingly complex in terms of the numbers and kinds of suppliers, the types of weapons available for purchase, and the nature of the transfer arrangements. In the early 1950s, a small number of countries – most notably the United States and the Soviet Union – transferred relatively limited amounts of mainly second-grade equipment to selected Third World clients. By the end of the 1980s, however, there was a dramatic increase in the number of states wishing to import modern weapons systems. In addition, Third World suppliers emerged as alternatives to the superpowers and the industrialized world. These Third World countries developed military–industrial infrastructures capable of producing sophisticated weapons systems, which, for commercial, political or technological reasons, they wished to market. The diffusion of supply sources was extended by the appearance on the international arms market of private companies and individuals willing to transfer expertise for commercial gain. In many instances, non-state suppliers functioned outside the boundaries of state law. The ability of second-tier and non-state suppliers to provide high quality equipment has dramatically improved Third World access to sophisticated weapons technology. By the end of the 1980s, ballistic missile supply had become one of the most prominent indicators of the changing patterns of international weapons transfers.

Three main categories of suppliers are crucial to the proliferation of ballistic missiles and related technology. First, there are the major state suppliers, which comprise three sub-groups: the superpowers, Europe and the PRC. These suppliers are primarily motivated by military advantage, political benefit and financial reward. In the second category of so-called second-tier Third World suppliers, two kinds of states stand out: those for whom export considerations are more important than domestic deployment and those who engage in missile technology transfers, but for whom domestic security considerations far outweigh the economic benefits deriving from sales. Finally, there are two types of non-state suppliers. There are a small number of companies and consortia which have developed specifically for the purpose of aiding Third World countries interested in acquiring missile and space launch systems. More commonly, regular businesses either knowingly or unwittingly participate in the proliferation of missiles and related technologies. The motivations of these non-state suppliers are financial.

MAJOR POWER SUPPLIERS

Since the 1960s, the most active missile supplier to the developing nations has been the Soviet Union. Arms exports have long been one of Moscow's most important means of demonstrating support for its Third World clients and a primary source of foreign currency earnings. Furthermore, arms sales have proved vital in maintaining economies of scale in Soviet defence industries, as well as providing an outlet for obsolescent equipment. Moscow has consequently sought to retain a major share of the international arms transfer market to the Third World in weapons in general and, until recently, in missiles in particular.[57]

In fact, according to one missile analyst: 'The current problems with ballistic missile proliferation in the Third World were almost entirely provoked by the indiscriminate Soviet sales of ballistic missiles to their Middle East clients. Nearly all the ballistic missiles in conflicts since 1945 have been Soviet manufactured missiles'.[58] Of the 22 countries under review in this study, twelve (Cuba, Afghanistan, North Korea, Iraq, Egypt, Iran, Syria, Libya, South Yemen, Kuwait, North Yemen and Algeria) have Soviet systems. More specifically, *FROG*-4s were supplied to Cuba in 1961; *FROG*-7s to Egypt in 1968; to Iraq and North Korea in 1969; to Syria in 1973; to Algeria in 1975; to Libya in 1978; to South Yemen in 1979; and to Kuwait in the late 1970s; *Scud* Bs were transferred to Egypt in 1973; to Iraq in 1974 (and again in the mid-1980s); Syria in 1974; Libya in 1976; South Yemen in 1979 and Afghanistan in the late 1980s; SS-21s were provided to Syria in 1983, and South Yemen and North Yemen in 1988. Significantly, the greatest single transfer of ballistic missiles to the Third World since World War II was at Moscow's behest: approximately 1,000 *Scuds* were sent to Afghanistan over the course of the 1980s. Iran received its Soviet-built *Scuds* from Syria, Libya and North Korea. North Korea is thought to have received *Scuds* from Egypt in the mid-1970s.

This does not mean that the USSR has dealt with missile transfers in a manner similar to other weapons systems. During the 1980s, when financial remuneration rather than support of political alliances appeared to have become the main motivating factor in Soviet arms transfers to the Third World, there was no significant increase in the number of states acquiring Soviet missile systems. During the 1960s and 1970s, Cuba, North Korea, Libya, Syria, Algeria, Egypt, South Yemen and Iraq received Soviet *FROG* and/or *Scud* systems. These states represented a substantial number of Moscow's Third World ideological and political allies. In the 1980s, when the number of clients for Soviet weapons was on the increase, only three new states, Kuwait, Afghanistan and North Yemen, joined the club. Although the USSR improved the quality of its missile exports in the 1980s, only Syria and North and South Yemen have been able to purchase the relatively accurate SS-21. Moreover, the Soviet Union has refused to provide Damascus with SS-23s or Libya with advanced missiles, such as SS-21s or SS-12s. Despite the certainty that a strong market for ballistic systems exists in both Iran and Iraq (following its transfer of *Scuds* to Iraq in 1987), the Soviet Union has demonstrated restraint in supplying either country with missiles. Indeed, since 1988 the only states which Moscow provided with missile systems were North and South Yemen and Afghanistan.

It is possible that Soviet missile transfers would have become far more extensive had it not been for Moscow's growing desire to reach some sort of accommodation with the West *vis-à-vis* ballistic missile arms-control proposals. In addition, the Soviet Union's recognition of its own vulnerability to missile attacks and a fuller appreciation of the implications of missile proliferation for regional stability may have influenced its missile transfer policy.

The United States has also become increasingly sensitive to the problems of ballistic missile proliferation. Indeed, it has been much more circumspect than the USSR about supplying complete ballistic missile systems to Third World states. Since the 1970s, Washington has viewed missiles as an extremely dangerous class of weapon, capable of delivering warheads of mass destruction and potentially damaging to regional stability and international security. As part of its policy of boosting the defence of its regional allies, the US provided South Korea in the late 1950s and Taiwan in the early 1960s with *Honest John* missiles, but it has studiously avoided extending the number of Third World states to which it supplies such weapons. Nor has it sought to upgrade the missile systems of South Korea or Taiwan.

The only other case of a country outside of NATO acquiring US ballistic missile systems is that of Israel, which in the 1970s received *Lance* missiles. However, the Israeli request to acquire the more sophisticated *Pershing* ballistic system was refused by the Americans – this despite the very close political relationship between the two countries.

In the past, however, American circumspection with regard to ballistic missile transfers was not accompanied by sensitivity towards the spread of space rocket launchers. As noted earlier, the development of space launcher technology has been an important means by which Third World states can acquire ballistic missiles. From the late 1950s, the US began transferring space technology (including suborbital sounding rockets) to about a dozen Third World states, including Argentina, India and Brazil. The American policy was motivated by the need to attract Third World states into the Western camp by offering them the best of Western technology, and the belief that technology transfers to the developing world would help promote international peace. This approach shifted as the relationship of space to missile technologies became more apparent. Then, the US took the international lead in demanding greater restrictions in the area of space technology transfers. Since the 1980s, US missile and space technology transfer policies have been better co-ordinated. Indeed, it has been the United States which has taken the lead within the international community in developing a global arms-control mechanism for ballistic missile proliferation.

The PRC has not shown the same self-restraint. Instead, Beijing has followed an aggressive arms sales policy aimed at both securing the hard currency necessary for its military modernization schemes and at challenging Soviet and American influence in the Third World. Chinese leaders see missile sales as an important part of this strategy and, indeed, CIA Director William Webster was adamant that the PRC was 'actively promoting' the export of shorter-range ballistic missiles.[59] So far, Beijing's main ballistic missile exports have been to Saudi Arabia, which purchased the CSS-2, the first long-range system ever to be exported to a Third World state.[60] The PRC is also thought to be aiding Iran in that country's ballistic missile and rocket programmes – specifically, in helping to develop the *Oghab* system as well as assisting in the establishment of a *Scud* B production line.[61] The PRC has recently agreed to help Egypt in its *Scud* development and other missile projects.[62] However, PRC Prime Minister Li Peng made a vague promise to President Bush in 1989 that the PRC would 'act responsibly on missile sales'.[63] To date, despite continued reports of impending Chinese missile sales to Libya and Syria, missile transfers to these countries do not seem to have taken place.

The PRC also seems interested in becoming one of the leading suppliers of space launcher technology to the Third World. The most extensive space agreements appear to be with the Brazilians, agreements which have included the establishment of a joint satellite launch company with Beijing.[64] The Chinese have yet to acknowledge the inherent dangers which the transfer of space technologies poses for ballistic missile proliferation.

None of the European states except France have developed ballistic missile capabilities, nor do they possess, aside from France and the UK, a significant share of the Third World arms supply market. Consequently, European involvement in the transfer of missiles and missile component parts has been relatively small. The only instance of a comprehensive West European transfer of missile technology and expertise to a Third World state is that of French assistance in the Israeli *Jericho* project during the 1960s. Support of Israel was viewed as crucial to French Middle Eastern policy at the time, and a range of co-operative agreements was undertaken between the two countries.[65] These were cancelled after the 1967 Six Day War, when French policy shifted from its pro-Israeli stance. Since then, France has refrained from complete missile transfers, although during the late 1980s it actively sought to export rocket technology to Brazil and India for their space programmes, even though this could be used in Brazilian and Indian missile programmes.[66]

Most criticism of European governmental approaches to ballistic missile proliferation has focused not on interstate activities, but on permissive government policies which have enabled companies to export technologies for use in Third World missile programmes. The United Kingdom, Italy, the FRG and Sweden are among the European countries accused of having allowed – albeit unwittingly – private companies to transfer equipment to Iraq which could be employed in Baghdad's military missile programmes.[67]

In Eastern Europe, the activities of the German Democratic Republic (GDR) were a concern for many years. In September 1988, a GDR delegation visited Tripoli to discuss bilateral co-operation in the area of ballistic missile proliferation.[68] In the past, GDR personnel may have been involved in both Egyptian and Iraqi projects. However, the recent political changes in the GDR will undoubtedly affect projects of this nature as East Germany's policies fall in line with those of the FRG.

SECOND-TIER SUPPLIERS

One of the major shifts in the pattern of arms supplies to the developing nations in the past decade has been the entry of Third World weapons manufacturers into the market.[69]

These countries are primarily attracted to the weapons export option by the potential economic rewards. Countries such as Brazil, Egypt and Israel regard weapons sales as a major source of foreign currency for their depleted treasuries. Many second-tier suppliers have few domestic products capable of bringing in the same amount of foreign exchange. In addition, domestic military requirements may be too small to allow economies of scale: arms exports can help make production costs easier to bear and domestic acquisition of various weapons more economical. In addition, as with the major powers, second-tier suppliers are not uninterested in securing political and military influence in recipient states. Although by superpower standards such second-tier ambitions may be modest, influence gained from weapons transfers can serve to boost a developing country's regional and international position. The ability to supply – or withold supplies – may also result in greater bargaining leverage and diplomatic benefits for the supplier. Finally, states may simply exchange weapons for technologies and expertise which they could otherwise not obtain. Technological barter is one way of bypassing external restraints on technology transfers.

Third World ballistic missile and technology transfers are highly significant as these weapons often represent the peak of Third World military technology and are perceived to have major strategic implications, both regionally and internationally. Furthermore, as demonstrated by the Israeli–South African, Syrian–Iranian and Libyan–Iranian cases, Third World states may act as suppliers of missiles or missile technologies to countries isolated by Western suppliers. Major power restraints on missile transfers can consequently be undermined.

Two cases that demonstrate the problems posed for missile proliferation by the rise of second-tier suppliers are Brazil and Israel. With only 10% of domestic production being taken up by national defence forces, the Brazilian defence industry remains highly dependent on exports. Although its arms sales have recently decreased, Brazil was, between 1985 and 1989, the ninth-largest exporter of weapons to the developing world.[70] Underlying this success was the relative cheapness of Brazilian weapons, their simplicity in design and ease of operation. But perhaps of greatest importance was the absence of end-user clauses. As one Brazilian defence brochure put it: 'Brazil abides by the principle that the sale of defence material is not a part nor an instrument of Brazil's foreign policy. No restrictions regarding the re-export of military equipment are imposed by Brazil because it believes that any possible decision in this area pertains to the sovereignty of the purchasing country'.[71]

Brazil does not publicly differentiate between the export of missiles and other weapons systems. It has already exported rockets (the *Astros* system based on *Sonda* technology) to Iraq.[72] Nor does Brasilia appear to be excessively concerned about the reliability and political stability of its potential clients. At the beginning of 1988, as part of a wide-ranging weapons deal, Libya offered to provide funding for a Brazilian missile, provided it would get the first batch of MB/EE missiles to be produced and would be guaranteed access to other Brazilian missile technology.[73] Although the US exerted pressure to squash the Brazilian–Libyan deal, Brazil's ambassador to Washington rejected criticisms of the agreement and noted that the Brazilian economy 'does not allow us to reject any deals'.[74] However, much to Brazil's displeasure, American and international pressure forced it to halt the sale.

The decline of the Brazilian arms industry (brought about by the termination of the Iran–Iraq War and a general decrease in foreign demand for weapons) may ultimately lead Brazilian exporters to promote weapons for which there is a growing market, such as missiles. The final balance between these pressures and external demands for missile supply restraints is difficult to determine.

Israel is another country heavily dependent upon weapons sales to sustain its large and sophisticated arms industry. Although its arms sales have recently begun to decline, Israel remains one of the developing world's most important arms suppliers. In the field of missiles and missile technologies, however, Israel has been circumspect. This stems from a wish to maintain the level of secrecy surrounding its own systems, from its recognition of the dangers of missile proliferation, and in part from a dearth of suitable clients. Indeed, aside from 'pariah states' such as South Africa and Taiwan, there are few to whom missile technology could be transferred without the risk of it ending up in the Arab world.

Israel probably engages in missile technology transfers to boost its ailing economy, to secure enriched uranium from South Africa for nuclear weapons, and for information and technology exchanges with Pretoria.[75] Israel's reputed willingness to assist South Africa's domestic missile programme may also grow out of its desire to make use of the South African coastline as a missile test range, away from the prying eyes of Arab states. It may also want to launch satellites in line with the earth's rotation; this is not a possibility in the Middle East unless Israel risks its test missiles landing in the territory of its Arab neighbours.

Israeli policy-makers vehemently deny that there is a missile co-operation agreement with Pretoria, stating that no new agreements have been signed with South Africa since 1987. The US State Department has maintained that there is no indication that Israel has sent American missile technology to South Africa, but the problem of verification is complicated by the fact that experience gained in the use of shorter-range missiles (such as the American *Lance*) could be used to construct longer-range systems.[76] This, of course, is not a problem unique to Israel; it is inherent in any technology transfer to the Third World.

The potential danger from second-tier suppliers thus appears small: few states have the technological capabilities of Brazil, while Israel has few potential clients and seems to be highly circumspect about missile transfers. On the other hand, as noted, there are a number of cases where ballistic missiles (or modified versions of these weapons) procured by Third World states from the major powers were subsequently transferred to other Third World countries. Examples of such missile exchanges include those between Egypt and North Korea;[77] Libya and Iran;[78] Syria and Iran;[79] and North Korea and Iran.[80]

Technology transfers and personnel exchanges have also taken place between Third World states. At one stage, the *Condor* 2 missile programme involved Iraqi, Egyptian and Argentinian co-operation in both of these areas. North Korea is one Third World country heavily involved in the transfer of manpower and expertise; North Korean personnel may be helping Egypt's domestic *Scud* production programme and the Iranian *Oghab* rocket project.[81] Israel may have assisted the PRC with its guidance programme. Brazil is reported to have aided in training Iraqi and Saudi Arabian missile personnel.[82] In short, for second-tier suppliers, the financial, military and political incentives to engage in missile technology transfers are limited only by constraints on technology and, in some cases, external political pressures. Certainly, technological constraints may be overcome, albeit slowly, as technological and military industrial bases expand; pressures for restraint can, in many instances, be neutralized through clandestine actions.

NON-STATE SUPPLIERS

The third group of missile technology suppliers includes individuals and business concerns actively engaged in providing Third World states with missile guidance, control, propulsion and testing systems. The most significant of these are individuals and businesses whose main activity is the provision of missile technologies to the Third World. Their emergence as suppliers has been made possible by the growing number of personnel in the developed world with the necessary technical knowledge to construct missile systems, many of whom are prepared to build missiles for financial gain. In addition, the limited ability of state law and authority to control such activities, the growth in the number of potential procurers and the large sums that procurers are prepared to pay for missile capabilities have contributed to the rise of non-state missile suppliers.

The earliest example of individuals coming together to construct a missile system for a Third World state took place in the 1960s, when a group of former Nazi engineers was employed by President Nasser of Egypt to construct a weapon similar to a V-2 rocket. The operation was soon aborted due to technical difficulties and attacks by the Israeli secret service against the German engineers involved in the project.

In the 1970s and 1980s, the FRG's tendency to ignore the foreign trade activities of its citizens allowed a number of German companies to become major proliferators of missile technologies. In the 1970s, the Orbitale Träger und Raketen Aktien Gesellschaft (OTRAG) signed an agreement with Zaire, leasing territory to develop a cheap launch vehicle capable of capturing a potential Third World space launch market and, probably, the military missile market as well.[83] External insistence, not least Soviet complaints which maintained that OTRAG was developing a weapon for Bonn, and ultimately German pressure, led to the closure of the Zairean site. France and the US also strongly opposed the project. An agreement was later reached between the company and Mu'ammer Gaddafi, who reportedly provided a missile test site at Libya's Siwa Oasis and large sums of money in exchange for missile technology. As a result once more of German pressure, OTRAG was soon forced to split up, but not before testing a prototype rocket.[84]

A far more effective organization is the Consen group, a consortium whose activities have been associated with the *Condor* 2 missile. In mid-1987, a sister company of Consen, *Condor* Projek AG of Zug, was alleged to have signed a contract with the organization believed to be running Iraq's missile programme, the Technical Corp for Special Projects, whereby it agreed to act as a technology consultant to Baghdad's missile development project.[85] Consen sent Iraq and Argentina technical personnel, many of whom had experience in European projects, including the European space rocket *Ariane*.[86] In 1988 one company in the Consen group, the Ifat Corporation, was reported to have helped finance an attempt to procure missile technology in the US, while in 1984 another Consen company, Desintec, allegedly made an illegal attempt to secure rocket nozzles for the *Condor* project from a Californian company.[87] Even more significantly, the Consen group is alleged to have encouraged and sub-contracted a wide variety of established European defence firms to contravene (knowingly or unknowingly) export laws and transfer vital technology to Iraq, particularly before the Missile Technology Control Regime (MTCR) came into effect in 1987.[88] (The MTCR will be discussed in detail in Chapter III.) Consen certainly represents a more sophisticated type of organization than OTRAG.

The emergence of non-state suppliers complicates both the supply process and potential arms-control regimes. Indeed, a significant amount of MTCR effort has been directed at restricting Consen activities.

Conclusion

Over the past few years, a number of developing states have attempted to demonstrate greater political and military autonomy. Consequently, they have also sought to acquire a range of increasingly sophisticated weapons systems. Indeed, the spread of ballistic missiles is but one indication of the diffusion of military and technological power within the international system. Certainly the explanations put forward by Third World proliferators for missile acquisition reflect their recognition of both the political and military advantages associated with the possession of ballistic missile systems.

The processes by which missiles have been acquired are diverse, although with the increase in demand and the onset of greater supplier restraints, domestic production is steadily taking over from procurement as the main source of supply. This has been made possible by the improved technology and scientific capabilities of the developing countries. However, in most cases indigenous production remains vitally dependent on external assistance. At the same time, economic motivations are becoming increasingly important and are replacing military and political motivations for *supplying* missile systems and technologies. This has taken place as the major powers have begun to withdraw from the market and be replaced by second-tier suppliers, whose military and political goals are much less ambitious. As a result the transfer of complete systems is giving way to the transfer of technologies and manpower, which are more in line with the capabilities of Third World and company suppliers. As far as ordnance is concerned, despite the proclivity of Third World states to play down the ballistic and nuclear/chemical linkage, there is a strong correlation between these types of proliferation. The question remains whether the spread of missiles is reinforcing an interest in non-conventional ordnance, or stimulating interest where none exists. Unfortunately, the level of secrecy surrounding proliferation decisions and the long lead times of weapons projects make any causal relationship difficult to determine. It may be argued that North Korea's nuclear research may only have followed its missile research and that Arab chemical weapons research may have been given added momentum by Israel's nuclear programme. However, it is far from certain whether missiles *per se* within the Third World (or even chemical and nuclear programmes) have served as major forces behind other non-conventional proliferation projects.

An analysis of the declared motives of proliferators, the processes of proliferation, and estimates of numbers and capabilities of systems highlights important parts of the ballistic missile proliferation problem. It reveals some of the perceptions of proliferators and their motivations for acquiring ballistic missile systems. In addition, it clarifies the scope of the phenomenon in terms of its most intense national and regional expressions. However, this first step cannot determine whether or not ballistic missile proliferation has adverse security implications. Nor can it indicate whether or not the issue of ballistic missile proliferation warrants the attention of arms controllers. In order to reach such conclusions, it is necessary to analyse the spread of ballistic missile systems in terms of its implications for regional conflict and international security.

Third World Ballistic Missiles

(1) Third World Long-Range Missiles and Space Launch Vehicles (SLV) > 900 km

Country	System	Status	Origin	Range (km)	Payload (kg)
Argentina	*Condor* 2	development suspended?	Joint Arg/Iraqi/Egyptian project with technical assistance from French, FRG, Italian, Swedish, Swiss sources[b]	1,000[a]	450–500
Brazil	SS-1000	dev planned	Brazil	1,000	not available
	VLS[c]	dev	Brazil	–	–
Egypt	*Badr* 2000	cancelled	Egyptian/Iraqi/Arg project, tech asst from USA, FRG sources	1,000	500
	Al-Ahred	canc in 1960s	FRG	950	n.a.
India	*Agni*	tested 1989	India with tech asst from FRG sources	2,500	680–900
	SLV-3[c]	tstd 1980	India with French tech asst		
	ASLV[c]	unsuccessful tests 1987/88	India	–	–
	PSLV[c]	under dev	India	–	–
	GSLV[c]	under dev	India	–	–
Indonesia	SLV[c]	planned	Indonesia	–	–
Iraq	*Al-Abbas*	deployed	Iraqi (modified Soviet *Scud*) with tech asst from Brazil and Egypt	900	500
	Condor 2	under dev?	Iraqi/Arg/Egyptian project with tech asst from Austrian, FRG Swedish, US sources	1,000	450–500
	Tammuz I	tstd 1989	Iraq	2,000	n.a.
	Al-Abid[c]	tstd 1989	Iraq	–	–
Israel	*Jericho* 2 B	tstd 1987	Israel	1,500	750
	Shavit[c]	tstd 1988/89	Israel	–	–
Pakistan	SLV[c]	planned	Pakistan	–	–
Saudi Arabia	DF-3 (CSS-2)	deplyd	PRC	2,200	2,000
South Africa	SLV[c]	planned?	SA with tech asst from Israeli, Taiwanese, FRG, French sources	–	–
South Korea	*Centaur*[c]	canc 1980	S. Korea, US design	–	-
	SLV[c]	planned	S. Korea	-	–
Taiwan	*Sky Horse*	canc 1981	Taiwan with US training	1,000	n.a.
	SLV[c]	under dev	Taiwan	–	–

[a] All figures must be regarded as approximates.
[b] Sources may refer to either government or private assistance. Other parties may have been involved in addition to those cited in this column.
[c] Space launch vehicle – surface-to-surface range approximates and payload not provided.

(2) Third World Medium-Range Missiles 200–900 km

Country	System	Status	Origin	Range (km)	Payload (kg)
Afghanistan	*Scud* B	dep & emp	USSR	280	1,000
Brazil	SS-300	suspended?	Brazil	300	n.a.
	MB/EE 350	under dev	Brazil	350	n.a.
	MB/EE 600	under dev	Brazil	600	n.a.
Egypt	*Scud* B	dep & emp	USSR	280	1,000
	Scud 100	under dev	Egypt with tech asst from PRC, N. Korea	n.a.	n.a.
	Al-Zafir	canc in 1960s	Egypt with tech asst from FRG sources	370	n.a.
	Al-Kahir	canc in 1960s	Egypt with asst from FRG sources	600	n.a.
India	*Privthi*	tstd 1988	India	250	1,000
Iran	*Scud* B	dep & emp	N. Korea, Libya, Syria	280	1,000
Iraq	*Scud* B	dep & emp	USSR	280	1,000
	Al-Husayn	dep & emp	Iraqi modified Svt *Scud*. Iraq with tech asst from Egypt, Brazil	600	140–250
	Fahd	under dev	Iraq[b]	500	450
	Baraq	under dev	Iraq[b]	250	n.a.
Israel	*Jericho* I	dep?	Israel with tech asst from France	480	250
	Jericho 2	dep?	Israel	750	450–680
Libya	*Scud* B	dep & emp	USSR	280	1,000
	OTRAG	under dev?	Libya with tech asst from FRG sources	500	–
	Al-Fatih	under dev?	Libya with tech asst from FRG	480–720	n.a.
North Korea	*Scud* B	dep	Missiles from Egypt, N. Korea reverse engineered	280	1,000
	Scud PiP	under dev	N. Korean modified *Scud* with tech asst from Egypt, Japan	500	n.a.
Pakistan	*Hatf* 2	tstd 1989	Pakistan with tech asst from France	300	500
	Hatf 3?	under dev	Pakistan	600	n.a.
South Yemen	*Scud* B	dep	USSR	280	1,000
South Africa	*Jericho* I?[d]	tstd 1989	Israel	480	250
South Korea	modified *Nike Hercules*	dep	S. Korea, modified US technology	240	n.a.
Syria	*Scud* B	dep	USSR	280	1,000
	OTRAG	canc 1981	Libya with tech asst from FRG sources	–	–

[d] Some sources claim the missile to be a *Jericho* 2.

(3) Third World Short-Range Missiles < 200 km

Country	System	Status	Origin	Range (km)	Payload (kg)
Algeria	*FROG*-4	deactivated	USSR	50	250
	FROG-7	dep	USSR	70	450
Argentina	*Condor* I	under dev	Arg, tech asst from Egypt, French, FRG, Iraqi, Italian, Swedish, Swiss sources	150	400
Brazil	MB/EE 150	under dev	Brazil	150	500
Cuba	*FROG*-4	dep	USSR	50	250
	FROG-7	dep	USSR	70	450
Egypt	*Sakr* 80	dep[e]	Egypt with tech asst from France	80	200
	FROG-5	dep	USSR	50	250
	FROG-7	dep & emp	USSR	70	450
India	*Devil*	canc in 1970s	India with tech asst from USSR	140	n.a.
Indonesia	RX-250	under dev	Indonesia with tech asst from France	100	n.a.
Iran	*Oghab*	dep & emp[e]	Iran, PRC	40	300
	Nazeat	dep	Iran, PRC	130	n.a.
	Shahin-2	under dev	Iran, PRC	110–130	n.a.
Iraq	*FROG*-7	dep & emp	USSR	70	450
	Laith	under dev	Iraq (modified Svt *FROG*)	90	n.a.
	Nissan	under dev	Iraq	110	n.a.
	Kassir	under dev	Iraq	150	n.a.
Israel	*Lance*	dep	USA	120	200
	Flower Project	canc in 1970s	Israel	200	n.a.
Kuwait	*FROG*-7	dep	USSR	70	450
Libya	*FROG*-7	dep	USSR	70	450
North Korea	*FROG*-5	dep	USSR	50	250
	FROG-7	dep	USSR	70	450
North Yemen	SS-21	dep	USSR	120	250
Pakistan	*Hatf* I	tstd 1989	Pak with tech asst from French sources	80	500
South Korea	*Honest John*	dep	US	37	580
South Yemen	*FROG*-7	dep	USSR	70	450
	SS-21	dep	USSR	120	250
Syria	*FROG*-7	dep & emp	USSR	70	450
	SS-21	dep	USSR	120	250
Taiwan	*Honest John*	dep	USA	37	580
	Ching Feng	dep	Taiwan, Israel	120	n.a.

[e] May be classified as artillery rockets. Other countries with artillery rockets are Brazil, Iraq, Israel and Saudi Arabia.

Data derived from multiple sources: Congressional Research Service Report, 'Missile Proliferation Survey of Energy Missile Forces', 3 October 1988; 9 February 1989; *The Military Balance 1989–1990*, (London: Brassey's for the IISS, 1989); *The Middle East Military Balance 1988–1989*, (Boulder CO: Westview Press; Tel Aviv: Jaffee Center for Strategic Studies); *World Military Expenditure and Arms Transfers 1988*, (Washington DC: US Arms Control and Disarmament Agency, 1989); Aaron Karp, 'Ballistic Missile Proliferation', *SIPRI Yearbook 1990: World Armaments and Disarmament* (Oxford: OUP, SIPRI, 1990); Table in *Arms Control Today*, May 1990, p. 31.

II. THE STRATEGIC IMPLICATIONS OF MISSILE PROLIFERATION

The central strategic questions associated with the spread of ballistic missiles in the Third World revolve around the impact of these systems on crisis stability and, in the event of conflict, on the processes of conflict escalation and intra-war deterrence. With crisis stability defined as a situation 'in which neither side can expect a lasting profit by actually initiating war',[1] the concern is whether a weapons system that guarantees rapid penetration of enemy defences and is capable of delivering both conventional and non-conventional ordnance undermines or supports Third World deterrent postures. It is necessary to identify the circumstances in which missiles could be – or are perceived to be – *decisive* weapons. At the same time, it is essential to assess whether, in the course of a conflict, the possession of missile capabilities – either mutually or unilaterally – will induce protagonists to widen and intensify the struggle, or impose restraints for fear of escalation.

The second set of strategic concerns regarding ballistic missile proliferation in the developing world derives from what may be termed the 'widening circumference of conflict'. The long-range capabilities of many Third World missiles under development ensures that the strategic implications of missile proliferation are no longer confined to a specific region, or even to the Third World. Both developing and industrialized states now – or will soon – have to contend with a new range of threats from sources previously regarded as either too far away to be strategically significant or incapable of penetrating air defence systems. These emerging threats have forced both the Third World and the industrialized states to reappraise their security policies and analyses of various defensive options.

If a specific weapon is militarily decisive, or is perceived to be so, decisions to initiate conflicts and the conduct of conflicts could be affected in disturbing ways. However, the military significance of such a weapon does not derive solely from its technical characteristics, but also from its military effectiveness within a broader operational setting. In some circumstances a weapon may be militarily decisive while relatively ineffective in others; a weapon can be a source of instability in some settings, but a stabilizing force in others. In some cases, the strategic implications of ballistic missile proliferation are simply muddled. Thus, an assessment of the strategic implications of ballistic missiles cannot be divorced from an analysis of their role in particular political and military settings.

Operational experience

Ballistic missiles have rarely been used in the heat of battle. Nevertheless, some lessons can be drawn from past operational experiences. These could serve to further our understanding of the strategic implications of the introduction and employment of ballistic missiles in the Third World. To be more specific, there are lessons to be learned about the role missiles have played in reinforcing crisis instability prior to the initiation of hostilities; the targets against which missiles have been used; the implications of missile employment for the type and scale of escalation; and the type of ordnance which has been delivered by ballistic missiles.

Since 1944 there have been only three sizeable ballistic missile campaigns: the German use of approximately 5,000 V-2s against the Allies in 1944–5; the Iranian–Iraqi missile exchanges between 1980–8, in which hundreds of missiles were launched; and the use by the Afghan government of approximately 1,000 *Scuds* against *Mujaheddin* forces during 1989–90.[2] Missiles have also been employed in two other instances, but on a much smaller scale. First, Egypt and Syria launched a small number of missiles against Israel in 1973.[3] Second, Libya attacked Lampedusa in 1986 with two *Scuds*.[4]

Concerns about crisis instability had little impact on the decisions to employ missiles in these campaigns. Germany and Iraq resorted to large-scale missile attacks against civilian targets relatively late in their respective wars, when they were both on the strategic defensive.[5] The Iraqis launched their major attacks on Teheran only in 1988, although they had used shorter-range *Scuds* in attacks against other population centres, such as Ahwaz and Dezful, earlier in the war.[6] The Afghan government used large numbers of *Scuds* against the *Mujaheddin* nine to ten years after fighting began; by the time these attacks took place, the Afghan government was on the strategic and tactical defensive.[7] The only case of missiles being employed in offensive operations at the outset of war was in the 1973 October War, when Syria and Egypt fired *FROG* missiles at Israeli command-and-control centres and airfields.[8]

The main objective in two of these cases was deliberate escalation of the conflict. Germany and Iraq used missiles to terrorize civilians on the other side of the front line in the hope of ending the war. On the other hand, Arab employment of missiles in 1973 was directed at military targets with the object of supporting general offensive military operations or later, attacking enemy bridges across the Suez Canal and military concentrations. Towards the end of the conflict Egypt fired a *Scud* missile into the Sinai desert. According to one analyst, the object 'was more political than military . . . suggest[ing] that Sadat would not allow the Egyptian army to be totally defeated without first using all of the weapons at his disposal.'[9] Gaddafi's attack in 1986 was aimed at the only American military facility within range of his missiles; he hoped to punish the US for its April 1986 raid on Libya.

Certainly, missiles have been more effective when they have been used against civilian rather than military or industrial targets.[10] The missiles launched by Egypt and Syria in 1973 against military targets and by Libya in 1986 missed their objectives and had no appreciable effect on enemy military performance or morale. The Afghan government's use of missiles against *Mujaheddin* forces besieging

Jalalabad coincided with the guerrillas' defeat. However, the part played by missiles in the government's successful defence of the city is uncertain. Because they are mobile and dispersed, guerrilla units are not easy targets for missiles. On the other hand, when directed at civilians in London during World War II and Teheran during the 1988 'war of the cities', missile attacks led to the flight of significant numbers of the population, which in the Iranian case resulted in widespread social and economic disruption. In Belgium and the UK, significant casualties occurred as a result of the German missile attacks. In the Iranian case, pressures resulting from the Iraqi missile salvos played a part in forcing the government to seek peace when it might otherwise not have done so.

The implications of missile employment for conflict escalation have been mixed. During World War II, there was no escalation in kind as neither the UK nor the US possessed surface-to-surface ballistic missiles. While attempts were made to destroy German launch sites, there was no increase in air attacks against German cities; extremely heavy bombing raids were already under way. Iran employed *Scuds* mainly in response to Iraqi air and missile attacks on its civilian centres. These attacks were supposed to emphasize countervalue targeting – Baghdad being the prime target – although Teheran claimed that it was firing at military targets in the city. The Iranian escalation to intensive attacks against cities might have dissuaded the Iraqis from continuing (though not completely abandoning) their aerial assaults on Iranian cities and industries.[11] Syrian *FROG* attacks against Israeli installations in 1973 were one of the factors which led Israel to bomb military and industrial targets deep inside Syria. Significantly, although Israel is said to have possessed *Jericho* missiles, it did not respond to the Syrian attacks in kind, but relied on its air force to carry out these deep penetration raids. Since Israel's air force maintained air superiority over Syrian cities, the Israeli government did not have to resort to missile attacks in order to retaliate against Syria. In any event, aircraft were for the Israelis more effective delivery vehicles than missiles. Finally, the United States did not retaliate following Gaddafi's attacks on Lampedusa. The ineffectiveness of the Libyan assault and the American desire to terminate the crisis as quickly as possible undoubtedly led to this decision.

In all five of these cases, only conventional warheads were used. One of the main reasons for this may have been the problem of developing chemical warheads for missiles and the technical problems associated with delivering a co-ordinated chemical-missile attack.[12] In addition, an element of self-deterrence might have been at work in both World War II and the Iran–Iraq case. Employment of chemical weapons (in World War II) or chemically-armed missiles (in the Iran–Iraq War) would have been a significant step in an escalation process which might have been difficult to control.

Four general points may be drawn from the history of missile employment. First, missiles played no role in the crises that preceded these conflicts. Second, missiles were more effective as weapons of terror against cities than as counterforce or counter-military systems. Third, the possession of missile capabilities by both sides did not automatically result in mutual deterrence. Fourth, the employment of missiles against a state with similar weapons did not automatically lead to a response in kind.

These patterns may or may not hold in the 1990s as ballistic missile proliferation continues, and as missile accuracies and payloads improve. As numbers of weapons expand, their potential for attacking significant numbers of targets will grow accordingly. As missile accuracy improves, the counterforce potential of missiles will rise; increases in missile payload and the ability to deliver non-conventional ordnance will raise the countervalue capabilities of missile systems exponentially. While in World War II, in past Arab–Israeli conflicts and the Iran–Iraq War, states could, prior to the beginning of hostilities, afford to ignore enemy missile arsenals, far larger and potentially more destructive arsenals may make pre-emptive attacks against missile forces appear necessary in the future. Should missile accuracies improve, pre-emptive counterforce attacks may be more practical than before. Conversely, while the possession of retaliatory capabilities has not deterred employment in the past, larger and more destructive missile arsenals may add to fears of escalation and reinforce pressures for intra-war deterrence. Thus, an examination of historical precedent can be of only limited value in a rapidly changing strategic environment. In the 1990s, the balance between stability and instability brought about by the proliferation of ballistic missiles will, as before, be highly dependent upon the specific political and military context in which the proliferation is taking place.

Missile proliferation and crisis stability in the Third World

A capability to guarantee rapid penetration of enemy defences and to attack military targets or civilian centres with either conventional or non-conventional warheads may undermine deterrent postures and, in a crisis situation, lead to the outbreak of war. However, it is important to recognize that ballistic missiles are not inherently destabilizing. Objective factors such as the balance of power and subjective elements such as governmental perceptions of the nature of the threat intrude to make strategic assessments dependent on the prevailing scenario. Thus, attention must be focused on existing balances of military forces and the broader military and political objectives of the parties involved.

To be more specific, when are there incentives to employ missiles as first-strike weapons against military targets? How often would pre-emptive counter-military attacks secure a favourable, if not a decisive, military outcome? The key factors in this regard are missile-to-target ratios, the necessity of destroying specific targets at the onset of conflict, and the importance of missiles – as opposed to other delivery systems – for securing these military objectives.

A survey of regions in which missiles have proliferated reveals only one example in which conventionally-armed missiles *may* have adverse implications for crisis stability: the Israeli–Syrian conflict on the Golan Heights. In other areas – Iraq and Iran in the Gulf, India and Pakistan in South Asia – conventionally-armed missiles are unlikely, either now or in the foreseeable future, to increase significantly the likelihood of war. It is only in the Syrian–Israeli case that the number of missile launchers deployed (in this instance by Syria) may cover the targets which would have to be neutralized in the earliest stages of a conflict.[13] In a surprise attack on Israeli positions on the Golan Heights, Syria would have to destroy Israeli defensive positions quickly and impede the Israeli mobilization process. According to one estimate, Syria possesses 24 *FROG* launchers, 18 *Scud* launchers and 18 SS-21 launchers.[14] If each SS-21 launcher is equipped with six missiles (that is, the number of missiles that could be fired before the launcher requires an overhaul), Syria could deliver between 36 and 54 tons of ordnance against Israeli targets. With six missiles to each *Scud* launcher, Syria could deliver 100 tons of ordnance with somewhat less accuracy.[15] When supported by 200 first-line attack aircraft (including MiG-23s, MiG-29s and Su-24s), another 800 tons of ordnance could be delivered during the first sortie.[16] Given the size of Israel's standing army in the north (two divisions), the Israeli target set is not extensive.[17] Virtually all of these targets are within range of Syrian systems. A Syrian military planner could therefore conclude that the majority of these Israeli installations could be targeted with at least some ordnance in an initial surprise attack.

Of course, missile attacks are not always successful. However, they do guarantee the penetration of enemy air defences and a good chance of inflicting some damage on vital targets. While aircraft are more effective than missiles in many respects (being more manoeuvrable, reusable, and having a greater payload capability than the missiles presently deployed in Third World arsenals), the Syrian Air Force is inferior to the Israelis', which means that missile capabilities would play an especially important role in Syrian first-strike calculations.[18]

However, one should be very careful *not* to overestimate the effects of conventionally-armed missiles of limited accuracy on armoured vehicles or hardened military emplacements.[19] Israeli air bases are among the best protected in the world – both actively and passively – and are supposed to function effectively even when under chemical attack.[20] Logistic and communication problems could also interfere with co-ordinated attacks, as would tactics undertaken to avoid Israeli counter-attacks. Certainly, the qualitative superiority of Israel's troops, air force and command-and-control system suggests that the Israeli Defence Forces' immediate response to a missile attack could be highly effective. In the future, Israeli anti-tactical ballistic missiles

(ATBMs) could destroy incoming missiles and reduce the impact of such an attack.

However, the manner in which threats are perceived may be of even greater importance than objective military considerations. Optimistic assessments by the Syrians regarding the ability of their missiles to destroy important Israeli targets could interact with pessimistic estimations from the Israeli side concerning the efficacy of a Syrian attack. The interchange of these two perceptions could lead to crisis instability. Deployment of missiles may thus bring about a situation in which 'the pressures are there for quick escalation during a crisis as it becomes a race in time between Syrian counter-military and Israeli counter-force attacks'.[21]

Indeed, that missiles could undermine crisis stability stems not only from Syrian incentives to strike first, but from Israeli incentives to prevent them from doing so. In the past, Israeli defence policy-makers have preferred to keep the initiative rather than absorb initial enemy attacks. While a willingness to absorb enemy attacks would be beneficial from the standpoint of crisis stability, the Israeli leadership does not believe it can tolerate the human and materiel costs involved. Israeli military and political leaders have certainly not sought to hide the fact that Israel might pre-empt Arab missile attacks. Pressures for pre-emption will become stronger should there be a significant increase in the number of missile systems at Syria's disposal, because this could change the missile-to-target ratio and increase Syrian first-strike capabilities even if missile accuracies remain unchanged. Syria has indicated an interest in expanding the size of its missile arsenal, and it may well do so in the not-too-distant future. Undoubtedly, a key factor from the standpoint of crisis stability is the missile-to-target ratio – a ratio that Israeli policy-makers must be watching with concern.[22]

In the Iran–Iraq and India–Pakistan cases, it seems less likely that missiles could serve vital strategic functions at the outset of a conflict. In these cases, therefore, deployment of conventionally-armed missiles would not undermine crisis stability. In neither the Gulf nor South Asia are military forces configured to launch surprise attacks – at least not to the degree that they are on the Syrian side of the Golan Heights. Moreover, in the Gulf and South Asia the ratio of missiles to military targets remains unfavourable, from the attackers' point of view. The large size of the armies, the length of the borders and the geographical depth of the countries involved must be juxtaposed against the limited amount of missiles already deployed or likely to be deployed in these areas. Missiles clearly do not pose a significant threat to enemy defensive positions in either of these regions.

In addition, missiles would be less important in pre-emptive counterforce or counter-military strikes, or in strategic attacks during the conduct of war, in South Asia and the Gulf than in the Syrian–Israeli case. In the former, air forces retain a greater offensive capability than in the case of the Syrians. Certainly, none of their enemies' air defence systems match the capability of the Israelis. In these cases, the fact that aircraft could be utilized for a variety of tasks, can be used repetitively and can deliver a large amount of ordnance makes them far more attractive than missiles.

Iran's air force, on the other hand, is aging and its F-4s and F-5s will find it increasingly difficult to penetrate Iraqi air space. Missiles could substitute for this deteriorating offensive capability. However, it is unlikely whether Iran (or Iraq, India or Pakistan) can readily develop missiles accurate enough to make reliable, conventional counter-military weapons.

The factors that make surprise counter-military missile attacks unattractive – the number of high value targets and missile inaccuracies – also make enemy preferences for pre-emptive action less likely. In addition, unlike Israel's past pre-emptive strikes, similar attacks in the Gulf and South Asia have seldom been particularly effective. Pakistan's opening strikes against the Indian Air Force in the early 1970s failed to destroy India's fighting capability. Likewise, Iraqi attempts to destroy the Iranian Air Force at the beginning of the Iran–Iraq War were unsuccessful. At present, missiles cannot guarantee that conventional pre-emptive attacks would be significantly more effective.

This does not mean that pre-emptive attacks are out of the question in South Asia and the Gulf. Should more missiles be deployed and system capabilities (especially accuracy) improve significantly, missile attacks would become more feasible. Some analysts have argued that by 2010 very accurate ring-laser gyroscopes and advanced inertial navigation systems could be available to a number of Third World states. Countries such as India and Israel could then deploy missiles with accuracies of approximately 30 metres – accuracies comparable to existing Soviet supplies.[23]

While missile deployments in most regions do not as yet seem to represent a significant military threat or a major motivation for adopting pre-emptive strategies, calculations are altered once the possibility of enemy missile strikes against population centres are taken into consideration. Indeed, as history has shown, it is in these instances that missiles can be highly effective, and it is here that many of the dangers of missile proliferation reside.

Escalation and countercity targeting

States may feel compelled to pre-empt not because they fear an enemy missile strike at the outset of conflict, but because they are concerned about the enemy's ability to escalate the war. Escalation could involve conventional or non-conventional missile attacks against civilian population centres. While the employment of conventionally-armed missiles against civilian populations would represent an escalation of hostilities, the use of chemical or nuclear attacks would raise the costs of a war enormously.

Although aircraft may be more efficient countercity weapons than missiles, Third World countries might still prefer to use missiles in attacks on enemy cities. Aircraft may be required for counter-military missions, while countercity targets are by nature less demanding from an accuracy standpoint. Second, missiles have a psychological impact that make them particularly effective terror weapons.

Third World states could employ missiles to attack cities, or to threaten to do so, for purposes of punishment, deterrence or compellance. The fear of missile attacks on civilian centres could deter a state from initiating or escalating a conflict. The employment, or threat of employment, of missiles against cities could compel an adversary to desist from military strategies he may otherwise have found profitable. Attacks on enemy civilians could be also be carried out for purposes of revenge.

Although non-conventional missile attacks against civilian populations would be particularly destructive, the prospect of conventional missile attacks against cities could also be destabilizing. Israel has always sought to keep its civilian casualties to a minimum and to protect its cities from conventional air attack; missile attacks are therefore high priority concerns. In Iran and Iraq, where there seems to be less sensitivity to civilian casualties, fears are nevertheless heightened by past experiences with conventional countercity missile exchanges and by the increases in the number and capabilities of missiles in the region. Should Iran succeed in its missile development plans, it would soon have enough weapons to attack all of Iraq's major cities. In mid-1988 analysts warned that:

> During the 'War of the Cities' the Iranians were able to fire only 77 *Scud* B missiles, that is, an average of only about one missile per day at Baghdad. By comparison, they now hope to be able to fire up to 20 *Iran*-130s per day. If the Iranians had possessed that number of missiles during the last 'War of the Cities', they would have been able to fire more than 1,000 missiles at Baghdad.[24]

On the other hand, because countercity attacks can cause great damage, missile proliferation need not be inherently destabilizing. Should missiles provide both sides with the ability to inflict heavy destruction on enemy population centres, then pressures to initiate conflict could be reduced. In the event of war, the dangers of countercity missile exchanges could compel the participants to impose restraints on the scope and intensity of the conflict.

That the spread of ballistic missiles could prove destabilizing in some circumstances and stabilizing in others is made more explicit in circumstances where missiles are configured to deliver non-conventional warheads. Then the dangers of initiating or escalating war leap considerably, as does the need to neutralize enemy capabilities as early as possible. Once again this is most clearly demonstrated in the Syrian–Israeli case.

Syria appears to have made a concerted effort to acquire chemical warheads for its missile systems.[25] While it is possible that Syria would employ its chemically-armed missiles against Israeli army and air bases at the initiation of hostilities – providing Syria with a greater chance of neutralizing Israeli forces during the crucial early stages of conflict – Syrian armed forces would then most likely have to bear the brunt of Israeli counter-attacks with chemical weapons. These attacks could impede Syrian progress across the Golan Heights. It therefore seems more likely that Damascus would seek to reserve employment of its chemically-armed missiles and either use them as a threat against Israeli attempts to escalate the war by attacking economic or command-and-control targets behind Syrian lines, or to prevent the total collapse of the Syrian Army. While the Arab states, including Syria, have publicly linked the possession of chemical capabilities to Israel's nuclear arsenal, the main role of chemical weapons could be to ensure intra-war deterrence in a conflict with Israel. Chemically-armed missiles could serve as a counter to Israeli air and army superiority. The Israeli Defence Force would almost certainly have to consider very carefully the risks of escalating a war through strategic bombing.

Should Syria launch missiles armed with chemical warheads against Israeli cities, Israel would feel obliged to respond in kind. Given Israel's air superiority, it is doubtful whether Israeli *Jerichos* would be used to deliver ordnance against Syrian civilian targets. As in the October War, Israel would probably rely on its air force for such missions. On the other hand, should Iraq enter the fray with missile attacks against Israel, the Israeli response could include missile attacks against Iraq – a necessity given the demands on the Israeli Air Force and the complexities associated with long-range air operations during a full-scale war.

In the Gulf, South Asia and South America, fears of enemy chemically-armed missile attacks against cities could reinforce stability and inhibit escalation should war break out. The case in point here is the Gulf, where both Iran and Iraq are deploying large numbers of missiles as well as extensive chemical arsenals.[26] Chemically-armed missile attacks against each other's cities, delivered by either missiles or aircraft, would most certainly cause massive casualties, and this would weigh heavily on the minds of Gulf policy-makers in any decision to initiate or escalate war.

During the Gulf War, Iraq demonstrated selectivity when employing chemical weapons. Chemical weapons were used against Iranian troop formations, but never against Iranian cities. The Iraqi Army mainly employed chemical weapons on the tactical defensive or as a means to disperse enemy troop concentrations.[27] Baghdad's use of chemical weapons against civilians (in Halabja against the Kurds) was made on Iraqi territory, against Iraqi citizens, and in the certainty that the Kurds could not respond in kind. Iran also confined its chemical-weapon attacks to military targets, with the exception again of Halabja. These self-imposed restraints may have reflected a recognition by the parties concerned of their mutual vulnerability to enemy chemical-weapon assaults and the desirability of strict limitations on chemical-weapon employment.

On the other hand, the fact that chemical weapons were not delivered by missiles may have been mainly a result of the technical problems associated with developing chemical warheads for missiles – a difficulty that Iraq may have since overcome. The Gulf is also characterized by intense hatreds, a propensity sometimes to take great risks, and a willingness to suffer what many in the West would regard as excessive casualties. Consequently, in a crisis or war situation in which one side faces defeat, restraints could rapidly evaporate. Indeed, '[t]he durability and robustness of a relationship based on mutual deterrence must surely be suspect in the context of such an unstable political environment as the Middle East.'[28]

It is important here to differentiate between the various kinds of non-conventional ordnance. While missiles armed with chemical warheads may cause far greater casualties than conventional missiles, should ballistic systems (or aircraft) be armed with nuclear ordnance, the potential for destruction will increase enormously. Missiles guarantee the penetration of nuclear ordnance into enemy territory and pressures to pre-empt in order to destroy nuclear-armed systems would be far greater than in an environment in which chemically-armed missiles are deployed. A precedent for attacking an enemy's nuclear facilities already exists. In 1981 Israel attacked and destroyed Iraq's *Osirak* nuclear reactor. Conversely, in other regions, the spread of nuclear ordnance and the ability to guarantee delivery may reinforce mutual deterrence. Indeed, in South Asia and South America there have been moves towards adopting bilateral confidence-building measures. Moderation appears to be spurred on not only by fear of nuclear countercity attacks, but also by concern about enemy attacks on nuclear facilities. Consequently, an agreement was reached in 1985 between President Zia of Pakistan and Prime Minister Gandhi of India not to attack each other's nuclear sites.[29] Confidence-building measures have also been adopted in South America between Brazil and Argentina, both of which have on-going launcher programmes and nuclear facilities not safeguarded by the International Atomic Energy Agency (IAEA). Thus, in 1987, Brazilian President Sarney visited Argentina's uranium enrichment plant and Argentina's President Alfonsín visited Brazil's Ipero plant.[30]

The Brazilian–Argentinian visits and the 1985 accord between India and Pakistan indicate that in these regions, at least, the proliferation of nuclear facilities may be acting as an incentive to reach mutually beneficial and stabilizing agreements. In the future, the further spread of ballistic missiles and non-conventional ordnance may reinforce such moves. Tentative steps towards confidence-building measures may reflect a recognition of shared vulnerabilities and the need to develop bilateral restraints. However, whether such accords will survive in severe crisis situations is uncertain. There is little experience with similar measures in South Asia or South America; in South Asia there are uncertainties about political/military control over missiles. For example, according to US government sources, Prime Minister Bhutto of Pakistan is not in control of her country's nuclear weapons programme.[31] This remains in the hands of the Pakistani Army. It is reasonable to assume that the delivery vehicles for nuclear weapons – which in the future may include ballistic missiles – will be out of her hands as well. And finally, there are problems of trust and difficulties associated with verification which could make adherence to agreements difficult in times of great tension.

Third World defensive options against ballistic missiles

Threats of second strikes against enemy targets may deter enemy missile attacks, but they will do little to reduce the level of damage should deterrence fail. Consequently, as missile proliferation continues, some Third World countries will probably seek some means of defence against missile attacks. There are four methods by which Third World states could guard against enemy missiles: *(1)* pre-emptive attacks; *(2)* ATBMs; *(3)* hardening and dispersion of important military assets; and *(4)* preparing the civilian population for the possibility of conventional and non-conventional missile attacks by developing a civil defence system.

While pre-emption is an obvious strategy, existing Third World operational capabilities may severely limit the effectiveness of pre-emptive attacks. The identification of enemy missile sites and the ability to carry out co-ordinated assaults that are both accurate and comprehensive requires command, control and intelligence capabilities that could be beyond the resources of most developing countries. Pre-emption also carries with it the dangers of escalation. Escalation could take the form of enemy missile attacks with surviving missiles or other military forces. Whether it is successful or not, pre-emption may result in high political costs, both domestically and internationally, which could render the option unattractive even if it is technically possible to execute. Thus, Third World countries could be drawn more to passive defensive measures.

One option is to shoot down incoming missiles with ATBMs. However, the ability to track and destroy incoming supersonic ballistic missiles is still a major technological challenge for states with sophisticated military and scientific industries. Not surprisingly, no ATBMs have been transferred to Third World states from external sources, and only two Third World countries – Iraq and Israel – are presently attempting to develop them. No ATBMs are yet deployed in the Third World.

In November 1988 Iraq tested an ATBM known as the *Faw* 1. Iraqi officials described the weapon as a response to Israeli missile capabilities, and heralded the successful test as a major demonstration of Iraqi technical prowess.[32] The Iraqi ATBM is probably a derivative of a surface-to-air missile. Whether, given the limitations of Iraqi military technology, it is a reliable and effective ATBM remains extremely doubtful.

A far more advanced ATBM system is the Israeli *Arrow* programme, which forms part of Israel's contribution to the United States' Strategic Defense Initiative (SDI). As such, Washington has promised to subsidize 80% of the development costs.[33] The sophistication of Israel's aerospace industry suggests that the *Arrow* could be a more effective ATBM than its Iraqi counterpart. Nevertheless, Israeli officials have refrained from sounding optimistic about the future of the *Arrow*. They have pointed out that it would have to be integrated with other defensive systems in order to be effective, and that deployment of the *Arrow* could cost up to $2 bn.[34] Should the system be deployed, it would probably be used at first to defend air fields and army bases rather than cities – a more limited and manageable task, but one that does not neutralize the dangers of enemy countercity attacks.[35] In the short term, therefore, ATBMs do not represent a viable option for most Third World countries.

Hardening military targets, including missile sites, to make them less vulnerable to enemy attacks, is a more realistic alternative for Third World states. It could also reinforce stability if developing nations acquired confidence in their capability to absorb an enemy first strike. Nevertheless, the costs of such an enterprise in terms of money and manpower would be considerable. While dispersal may be a way of obviating the financial and technical problems associated with hardening military targets, it is nevertheless an option which may appear unattractive to many Third World countries. Dispersal would inevitably lead to difficulties in command and control that could render missiles and other military capabilities less effective. Domestic political and military insecurities might make dispersal extremely dangerous, especially if the weapons are deployed in regions far from the command centre or by troops at odds with the central leadership.

Preparing the population for missile attacks through the construction of shelters, training in chemical protection and evacuation procedures also remains beyond the ability of most Third World states; Israel is the only obvious exception. There are a number of reasons for this. In the first place, it has a small population which is accustomed to a high level of military mobilization. Moreover, the Israeli civil defence system has been in place for decades. Consequently, an extensive shelter system, and tried-and-tested civil defence procedures are well established. No other Third World state has a comparable civil defence infrastructure. Large populations, difficulties in communications and the costs of construction and training make effective civil defence programmes very difficult to establish. In short, defensive systems or programmes are not attractive alternatives for most Third World states.

If introduced bilaterally (or multilaterally), any of these defensive strategies could reinforce deterrence and crisis stability. Defences would make first strikes and other counter-military attacks more difficult. Conversely, knowledge that damage to the homeland and military assets could be reduced may make states more willing to risk conflict. Should any of these defensive measures be introduced unilaterally, the impact on crisis stability will also be complex. Whether unilateral deployment of ATBMs, for example, would undermine crisis stability in the Third World would depend on the effectiveness of the systems in question and the political and military objectives of the state deploying the ATBM. ATBMs could act as a protective shield behind which a country could launch military attacks, secure in the knowledge that its homeland was protected. Clearly, there are a number of important similarities to be found amongst the arguments both for and against the deployment of Third World ATBM systems and those regarding deployment of the SDI. However, generalizations about the implications of ATBM deployment for stability remain highly scenario dependent.

The industrialized states and Third World ballistic missile proliferation

The proliferation of ballistic missiles in the Third World is a phenomenon that affects the security of the developed world as well. This in turn has forced the industrialized states to assess their defensive options with respect to Third World ballistic missile capabilities.

Ballistic missiles widen the circumference of a state's military reach and intermesh North and South security concerns to a greater degree. Thus, the Soviet Union could be targeted by Israeli, Iraqi, Iranian or Saudi Arabian missiles. The PRC could be targeted by Indian and Pakistani missiles. Eastern and Southern Europe could be attacked by missiles launched from Iraq, Israel or Saudi Arabia. Analysts in Turkey, which is a NATO member, have expressed concern about Arab missile capabilities.[36] While the United States is presently beyond the reach of Third World ballistic missile systems, its vulnerability would increase should countries such as Iraq, Brazil, Israel and India deploy long-range missiles. In any event, American bases overseas have become increasingly vulnerable to Third World ballistic missiles.

Of course it can be argued that Third World countries have long possessed the ability to attack the industrialized world with sophisticated combat aircraft. The major difference, however, rests in the penetration capability which missiles provide against states with well-developed air defence systems.

The psychological dimension of the problem is also of importance. Although it is possible to question both the efficacy of missiles and the readiness and likelihood of Third World states to use them against nations of the developed world, it is nonetheless almost certain that missiles will serve to change broad calculations of deterrence between both geographically contiguous and more distant adversaries. While it would be imprudent to overstate the dangers to NATO and WP countries, some proliferating states are politically unstable, possess insecure command-and-control structures, and have geographical and ideological disputes with NATO and the WP. In a crisis situation, it is not inconceivable that Third World countries could threaten to employ, or even actively employ, non-conventional ballistic missiles against states in the developed world.

It is therefore not surprising that the proliferation of ballistic missiles amongst Third World states is increasingly permeating discussions on Western security. These deliberations have touched upon a number of military, political and arms-control options. Pre-emption against Third World missile capabilities is more feasible for industrialized countries than for developing states. However, while the industrialized world may possess the necessary military capabilities for pre-emptive strikes, the international and domestic political repercussions of such attacks – as well as the military implications should the strikes fail – make direct military actions complex and in many cases, impracticable.

The possibility of deploying defensive systems against Third World ballistic missiles has also received growing attention. It has been pointed out that ballistic missile proliferation and future reductions in superpower nuclear missile arsenals as a result of the Strategic Arms Reduction Talks (START) negotiations will tend to increase the value of defensive systems.[37] For example, US Vice President Dan Quayle has recently reiterated that 'the proliferation of ballistic missiles is an important new development and lends weight to arguments favouring deploying an SDI system'.[38] Certainly, the declining Soviet military threat makes the Third World ballistic missile threat more prominent than ever and they are now often used as a partial legitimization for SDI deployment.[39]

Current American ATBM programmes involve research on the Extended Range Interceptor (ERINT) missile and the Theatre High Altitude Area Defence (THAAD) missile. The object of the former would be to intercept and destroy manoeuvering ballistic missiles while simultaneously delivering sufficient energy to destroy a chemical warhead before it could release its toxic agent. THAAD would destroy enemy missiles at much higher altitudes than ERINT.[40]

Before long the US will also have to decide whether to fund the next stage of the Israeli *Arrow* ATBM programme and begin preparations for production. The *Arrow* remains the ATBM closest to completion and, in the late 1990s, could be a vital component of both Israeli and US defences against Third World ballistic missiles.

Conclusion

The most significant strategic asset of ballistic missiles is their ability to penetrate enemy defensive systems. It is this capability that makes them attractive and, at the same time, dangerous. Because missiles may attack very distant targets with great rapidity and deliver conventional, chemical or nuclear ordnance, they are weapons of surprise attack that have great potential for destruction.

This is not to say that ballistic missiles will be militarily decisive in the future, or that they will necessarily be a source of regional instability in the 1990s. While conventionally-armed missiles may make extremely effective psychological terror weapons, there is only one area – the Golan Heights – where conventionally-armed missiles could help achieve meaningful military objectives; even here, the evidence is far from clear. During World War II Hitler hoped that missiles would be war-winning weapons. It was his belief that the V-1s and V-2s would create 'such a storm of protest and war-weariness that the [British] Government will be overthrown'.[41] However, both then and in more recent conflicts, conventional ballistic missiles have not been militarily decisive weapons.

Whether, as missile proliferation increases, missiles can increasingly promote the attainment of strategic objectives will depend on variables other than the quantity and quality of the missiles themselves. These variables include shifts in overall balances of forces, missile-to-target ratios, and the development of defensive systems.

One key variable that will affect the strategic ramifications of ballistic missile proliferation in the 1990s will be the ability of Third World states to manufacture missile warheads for non-conventional ordnance. Should chemical ordnance be married to missiles, fears of countercity damage and pressures for pre-emption will greatly increase. This process is already unfolding in the Middle East. If nuclear warheads are fixed to missiles, then the pressures to pre-empt in a crisis situation may become extremely pressing.

Thus, while missile and non-conventional weapons proliferation could serve a stabilizing function, there is no certainty that this will come about in regions long racked by mistrust, tension and acute disputes. However, it is important to recognize that the significant threshold here is not the deployment of nuclear-armed missiles as much as the successful development of nuclear weapons. Missiles would only serve to exacerbate a dangerous situation.

While the direct dangers to the industrialized world from Third World missiles should not be overstated, the proliferation of ballistic missiles is, theoretically, even more profound for these countries than the developing world. For, while until the 1980s, most Third World states have always been vulnerable to air attack from other Third World states, Europe and the Soviet Union have not. The proliferation of missiles changes this calculation, and should Third World proliferators develop non-conventional warheads for their missiles, it is inevitable that this will affect security planning in industrialized countries. Indeed, it is these concerns that have motivated Western attempts to impose controls on the proliferation of ballistic missiles.

III. PROBLEMS OF CONTROL*

Introduction

In the early 1980s, Western policy-makers began to develop a multilateral mechanism for controlling the proliferation of ballistic missiles. This new initiative took place against the backdrop of a growing international trade in missiles and an increased emphasis in a number of Third World countries on the research and manufacture of long-range ballistic systems and space launch vehicles. During the Iran–Iraq War, missiles were being fired in significant numbers for the first time since World War II. The fact that chemical weapons were also used during the Gulf conflict, together with fears that many of the new Third World ballistic missile types could be employed to deliver nuclear and chemical weapons, heightened concern about the linkage between non-conventional proliferation and the spread of ballistic missiles.

That a ballistic missile arms-control initiative was slow to emerge during the 1960s and 1970s was not surprising, given the small number of Third World domestic ballistic missile development programmes in progress at the time and the limited range of the missiles in the Third World. In addition, the linkage between nuclear, chemical and missile proliferation in the Third World had not received serious consideration in international arms-control forums. However, as the quantity and quality of Third World missiles increased, and fears of nuclear and chemical proliferation grew, the question of delivery vehicles in general, and missiles in particular, began to receive greater attention. The emerging centrepiece of attempts to contain ballistic missile proliferation was the Missile Technology Control Regime (MTCR). The MTCR began to be negotiated in secret between seven industrialized countries, (the US, the UK, the FRG, France, Italy, Canada and Japan) in 1983; the regime was publicly announced in 1987. Essentially an agreement between supplier nations, it aims to prohibit the transfer of nuclear-capable missiles to the Third World, and to restrict the supply of missile-related technologies.

The relatively short period which has elapsed since the creation of the MTCR, the complexity of the missile transfer network, and the limited membership and structural problems of the regime have meant that its success has been mixed. As a result, a number of proposals have been put forward either to modify the MTCR or to replace it with a broader and more ambitious agreement.

This Chapter outlines the context, scope and functioning of the MTCR. It should be noted, however, that much of the information on the origin and workings of the regime remains classified. This itself is testimony to the sensitivity of the Western industrialized states to the subject of missile proliferation and the controversial nature of the MTCR. The Chapter will also discuss some of the major difficulties of

* This Chapter was co-authored by M. Navias and D. Howlett.

the present regime with special reference to its limited international scope, the problems associated with dual-capable technologies, the bureaucratic obstacles which hamper the efficient implementation of export controls, and the need to take into account the security concerns of proliferating states. Despite these problems, it shall be argued that the MTCR is the most viable mechanism for controlling the spread of ballistic systems. It will also be contended that the regime should be viewed as a 'braking process' that seeks not so much to prevent totally the proliferation of missiles, but to slow the process down. The subsequent creation of a breathing space allows time for the introduction of political initiatives which could play a stabilizing role in areas of tension, thereby ameliorating the worst effects of ballistic missile proliferation. The merits of a supply side focus are then analysed and the difficulties of alternatives to the MTCR assessed. The Chapter concludes with a discussion of the measures being taken to address the weaknesses of the MTCR.

Background, scope and functioning of the MTCR

BACKGROUND

While little attention was paid to measures aimed at constraining the spread of ballistic missiles at the beginning of the missile proliferation era in the 1960s, it is possible to note, in the period prior to the MTCR negotiations, a gradual increase in interest, mainly within the US, in developing an international missile transfer control agreement. This growing concern was ultimately to find expression in the MTCR negotiations begun in 1983.

The issue of controlling the spread of missiles to the Third World was largely overlooked in the multilateral arms negotiations and treaties of the 1960s and 1970s. The NPT focused on nuclear ordnance rather than delivery systems, while American regional arms-control proposals did not distinguish between missile capabilities and other conventional arms.[1] This lack of concern was not surprising, given that Third World missiles were still few in number and relatively short range, and that Third World domestic production of ballistic missiles was still a decade away. Nevertheless, both the US and the Soviet Union appeared to observe tacit restraints when it came to providing clients in the Middle East and Korea with ballistic missiles.[2] The USSR restricted its supplies to limited amounts of 40–70-km *FROGS* and 280-km *Scud* Bs, while the US transferred only a small number of 37-km *Honest Johns* and 120-km *Lance* systems. Longer-range systems were not transferred, despite the demands of clients such as Israel. The US and the Soviet Union themselves did not acquire accurate tactical ballistic missile systems until the 1970s and 1980s. Nevertheless, there was an implicit recognition by both states that missiles could not be placed in the same category as other conventional arms transfers. Their long ranges, guaranteed penetration capability and destructive potential clearly indicated that missiles should not be freely transferred.

India's explosion of a nuclear device in 1974 stimulated attempts to move beyond these tacit restraints. The focus that emerged during the 1970s was not only upon the increasing need to constrain nuclear proliferation, but also to give serious consideration to the linkage between conventional arms and non-conventional proliferation. Concerns about continued nuclear proliferation led to the establishment in 1975 of the 'London Guidelines', which can be seen to have served as a model for the structure of the missile control regime. These guidelines did not constitute an international treaty, but rather an agreement between seven states (France, Canada, the FRG, Japan, the UK, US and the USSR) to adopt a common policy for nuclear exports, especially enrichment and reprocessing technology.[3] The supplier focus and the emphasis on self-restraint were to characterize the later approach towards restricting the transfer of missiles and related technologies. During this period the policy implications of the relationship of conventional weapons transfers to nuclear proliferation were, at first, not always clear. Indeed, some policy-makers believed that conventional arms transfers could be used as a means of dissuading clients from pursuing nuclear weapons and long-range missile options.[4] In the 1970s, for example, the United States helped persuade South Korea not to produce long-range missiles by providing Seoul with sophisticated military equipment, including F-5 fighter aircraft and production rights for American weapon systems.[5] Nevertheless, the view that if the recipient country decided to acquire nuclear weapons anyway, conventional arms transfers could help augment its nuclear ambitions, was also recognized, thus creating what has been called the 'dove's dilemma'.[6] These concerns helped persuade Washington to reject Israel's request for nuclear-capable *Pershing* 1A missiles.[7]

In the US the recognition that concerns about nuclear proliferation must be accompanied by concerns with regard to conventional delivery systems soon developed into a readiness to distinguish between delivery systems and other conventional arms transfers. For example, in a 1976 proposal to the Committee of the Conference on Disarmament, the US representative singled out missiles and manned aircraft for conventional arms-control purposes. He stated that the US restricted the supply of both aircraft and missiles in many regions. It was stressed that Washington opposed the transfer of systems which could cause large-scale damage to population centres. The US wished to persuade other suppliers to support such restraints and called on recipient states 'to forgo acquisition of destabilising systems not yet introduced in to the area, particularly surface-to-surface missile systems having a long range beyond any defensive need and aircraft having a long-range strike role'.[8]

The next major step towards establishing a missile control technology regime emerged during the late 1970s. The United States' willingness to differentiate between various types of delivery vehicles surfaced in the US Arms Control and Disarmament Agency's (ACDA) proposals for the Conventional Arms Transfer (CAT) talks, held with the Soviet Union. Between 1977 and 1979, ACDA sought to convince the Carter Administration that efforts should be made to distinguish between categories of delivery systems and equate them more explicitly with other proliferation concerns. It was ACDA's view that control strategies should focus first on those weapons systems that would not create too much controversy if global restrictions were placed upon them. As experience of transfer limitations grew, so the agenda would gradually move towards placing constraints on the more common items of the international arms trade.[9] This strategy meant that the first category of weapons for restraint would be the so-called 'weapons of ill-repute', which included long-range surface-to-surface missiles. While attempts to control specific categories of weapons gave way to an approach that stressed the primacy of regional political solutions, and although CAT was ultimately to be aborted, missiles had been singled out as a special case for arms transfer restraint.

The need for arms controllers to focus on missile systems was reinforced in the late 1970s and early 1980s by the continued proliferation of more capable ballistic systems, some of which could deliver nuclear ordnance, and by the emergence of domestic missile development programmes. The launch of the Indian SLV-3 in 1980 also increased US awareness of the dangers of proliferation in space technology. The problem of controlling the transfer of civilian technologies which could be employed for military purposes was becoming increasingly important as a number of states, such as India and Brazil, were utilizing their space technology for military research purposes.

Thus, by 1982, tacit American restraints on the transfer of missiles began to give way to efforts to develop domestic ballistic missile and space technology transfer policies and to co-ordinate these with other interested governments. In November 1982, the Reagan Administration adopted National Security Directive 70, which noted the problems posed by Third World missile proliferation and emphasized the need to deal with the issue. The first formal international discussions on controlling the spread of ballistic missiles took place between the US and the UK as a result of an American initiative in May 1982. In June 1982 a further meeting was held, this time including France, the FRG and Italy. The first full meeting, which took place in Washington in June 1983, also included Japan and Canada.[10] All the states, except Japan, were suppliers of conventional arms to the Third World and had previously transferred nuclear and space technology. The Soviet Union – the major supplier of ballistic missiles to the Third World – was excluded from the negotiations as it was believed that Moscow had little interest in constraining its missile transfers and would strongly oppose international attempts to develop an international ballistic missile arms-control regime. In such an environment it was thought best to secure Western agreement before making an approach to the Soviet Union.

In 1985 agreement was reached between the seven states regarding an international mechanism for controlling the spread of missiles and related technologies. While implementation of the regime began immediately, French attempts to win concessions from the UK and the US over the Treaty of Rarotonga – specifically that they would not sign the treaty for a nuclear-free zone in the South Pacific – delayed the formal public announcement of the MTCR until April 1987.[11] (In the United States this announcement was followed six months later by National Security Decision Memorandum-187, which prohibited the export of space launch technology on the grounds that it could further ballistic missile proliferation in the Third World.)

SCOPE

Neither an international treaty nor, in the technical sense, an agreement, the MTCR is essentially an attempt to establish identical sets of national export policies. Its aims are primarily twofold: Firstly it seeks:

> to limit the risks of nuclear proliferation by controlling transfers that could make a contribution to nuclear weapons delivery systems other than manned aircraft. The guidelines are not designed to impede legitimate national space programmes or international cop-operation in such programmes as long as such programmes could not contribute to nuclear weapons delivery systems.[12]

Secondly, it aims to ensure that all the major Western suppliers exercise similar types of export controls, thus preventing some supplier countries from benefiting from the restraint of others.[13]

To this end, the MTCR attempts to create a common evaluation process in assessing the export of relevant items. Two categories of items for control are defined. Category 1 consists of items which can only be employed for the purpose of constructing missiles. These are technologies of the greatest sensitivity and the MTCR is explicit that there must be 'a strong presumption to deny such transfers'. Category 1 includes complete rocket systems (including space launch vehicles, sounding rockets and ballistic missile systems) and unmanned air vehicles (including cruise missiles and drones), all of which are capable of delivering 500-kg payloads over 300 km. The payload limit was set in accordance with the minimum payload believed necessary to deliver a first generation nuclear warhead. The uncertainties surrounding the requirements for chemical-weapons delivery made the establishment of any chemical or biological criteria difficult – and indeed, such criteria did not enter the technical debate at all.[14] The 300-km range limit may have been fixed with the distance between Middle East capitals in mind.[15] Other items included in Category 1 were complete missile sub-systems, individual rocket stages, re-entry vehicles, liquid- or solid-fuelled rocket engines, guidance sets, thrust vector controls and warhead safing, arming, fusing and firing mechanisms. The transfer of production facilities for any of these Category 1 items was strictly forbidden.

MTCR constraints on exports are less stringent for Category 2 items. Many of these items are of a dual-use nature and may be used for purposes other than missile development or military missions. However, should there be any suspicion that these technologies are being procured for missile construction they will be denied. Category 2 items include: propulsion components, propellants, launch and ground support equipment, missile computers, analogue-to-digital converters, test facilities, structural materials, flight instruments, inertial navigational equipment and flight-control systems.

Thus the MTCR is a structure within which supplier governments must take responsibility for the end-use of any transferred technology. However, there is no central co-ordinating body to oversee adherence to MTCR guidelines, or to take punitive measures against countries, businesses or individuals contravening the agreement. Instead, the decision to transfer equipment remains the sole responsibility of the supplier government. It is the duty of the supplier to decide upon the suitability of the sale of any missile-related technologies, having assessed the purchaser's intentions and capabilities for producing nuclear weapons delivery systems other than manned aircraft.

The supplier must also take into account specific assurances by the recipient that the transferred technologies will not be used for missile purposes. The MTCR further maintains that another important factor in the supplier's decision to transfer missile-related technologies to a recipient is the existence of relevant multilateral agreements. While there is no specific reference to any particular multilateral agreement, the nuclear focus of the MTCR seems to indicate that there should be no transfers of missile-related technologies to countries which are not party to the NPT. In short, while ballistic missiles are treated as a distinct topic for the purposes of control, concern about this weapons system is linked to the issue of potential nuclear delivery. For example, in the accompanying UK statement to the MTCR, the concern was raised that missiles would exacerbate non-conventional proliferation. The UK's view was 'that the further spread of nuclear capable missiles could increase the risk of the proliferation of nuclear weapons themselves'.[16]

Finally, the signatories promised to 'exchange relevant information with other governments applying the same guidelines'. They called for all states to join them in adhering to MTCR guidelines.

FUNCTIONING

Central to the MTCR is the acquisition of intelligence on ballistic missile proliferation in the Third World, including the capabilities and intentions of actual or potential proliferators and, most importantly, their sources of supply for manpower, technical expertise and equipment. Should any of these originate in one of the MTCR member states, the relevant government is required to take legal steps against the individuals or companies involved. The full co-operation of the members with regard to data collection, assessment and collective action against those who transgress MTCR guidelines is crucial in this regard.

International co-operation and goodwill is vital to the efficient functioning of the regime, particularly in view of the increasingly complex and secretive transnational ballistic missile transfer market. Subsequently, in 1989, the Bush Administration opposed attempts in the Senate to pass legislation that would allow the US to act unilaterally against foreign firms or countries. As Assistant Secretary of State, Richard Clarke, told a Senate Foreign Relations Committee hearing, the strength of the MTCR is based upon 'diplomatic consultation and information exchange'. He went on to warn that any unilateral action by the United States:

> contravenes the spirit of the regime, and would very likely cause frictions that could seriously threaten the regime's effectiveness as a cooperative mechanism . . . the possibility that the US might act unilaterally against a foreign firm or country could seriously hamper the information-sharing mechanism of the MTCR. Partners who might wish to consult with us on missile-related activities of their own companies would be unlikely to do so if the result would be to invite US sanctions against those companies.[17]

The US has played a central role in setting up and co-ordinating the MTCR: it initiated the establishment of the regime, and has the greatest amount of resources to devote to it. In May 1989, the Director of the CIA informed a Senate Committee that the Agency had been collecting and assessing intelligence on the subject of ballistic missile proliferation for decades. He stressed that the CIA employs 'some of our most sophisticated collection methods and analytical techniques to address this critical issue'.[18]

According to Richard Clarke, the MTCR brings 'effective diplomatic pressure to bear – without the actual use of sanctions – to restrain the exports of other nations and retard the development of missile projects of concern'.[19] Deputy Secretary of Defense Hind was even more explicit when he stated that the MTCR worked 'through a combination of export controls, démarches to foreign governments and bilateral and multilateral discussions with our MTCR partners'.[20] Thus, a wide range of controls is employed, although in a manner which indicates a preference for quiet action rather than open threats. A great deal of effort has been directed towards the termination of the *Condor* 2 project, the deployment of which appears to have been postponed until the mid-1990s.[21] At a meeting of the MTCR governments in Rome in September 1988, *Condor* 2 was identified as a major 'project of concern'.[22] The fact that the weapon would be a nuclear- and chemical-capable TWLRM and that it would be deployed in the

Middle East and South Atlantic made it particularly dangerous. In addition, being a tripartite international project, dependent on private European expertise and parts, made it potentially vulnerable to outside interference. The fact that many of the private companies involved in the project originated in the FRG (an MTCR signatory) was another cause for concern. MTCR misgivings about the *Condor* were raised with the governments involved in the joint production programme and, subsequently, clandestine supply networks were blocked. Co-operative efforts with the German and Italian governments succeeded in preventing exports from private concerns based in those countries; in Italy and the FRG legal proceedings were brought against certain individuals, while in Bonn export laws and anti-proliferation legislation were tightened.[23] As a result of MTCR efforts, Egypt has withdrawn from the *Condor* 2 programme,[24] and Argentina has indicated that it would suspend its missile project.[25] A recent study claimed that Baghdad has also lost interest in the project.[26]

The *Condor* 2 has not been the sole focus of the MTCR. Following the PRC's sale of the CSS-2 to Saudi Arabia, discussions were held with Beijing, which has since undertaken not to transfer any more long-range missile systems to the Third World. In addition, both India and Brazil have complained that new export restrictions have affected their space launch programmes. India's difficulties in acquiring American missile-testing technology and French liquid-fuel technology have helped delay its rocket programme. New Delhi has recently indefinitely postponed its second test launch for its *Agni* missile. Brazil, in turn, has had to postpone the first launch of its VLS space launcher until 1992, while Avibras was forced to suspend its SS-300 ballistic missile programme.[27]

The limitations of the MTCR

While the MTCR can claim some successes in its struggle to contain the spread of missiles and related technologies, these appear relatively minor when viewed against the increase in missile proliferation and the qualitative improvements of many existing systems in the Third World. If anything, Third World domestic production of missile systems has increased since 1987 when the MTCR was made public. Third World ballistic missile systems with very long ranges, such as the Indian *Agni*, the Israeli *Jericho* 2 and the Iraqi *Tammuz*, were all tested after 1987. There has also been an expansion of several Third World space programmes, such as those of India, Brazil and Israel, which are possibly facilitating ballistic missile research. Consequently, there has been much criticism by Western arms controllers of the structure, objectives and functioning of the MTCR. These criticisms have generally focused on its limited international scope, the complexities surrounding the interpretation of its guidelines, and its inability to address the real security needs of Third World states.

THE NARROW INTERNATIONAL SCOPE

The assumption underlying the MTCR is that Third World missile programmes are vitally dependent on external technical assistance, yet the regime's greatest weakness is its slow progress in attracting a significant number of key technology suppliers.[28] At its inception, the MTCR was signed by four European, two North American and one Asian state. All were industrialized democracies and all, aside from Japan, were members of the North Atlantic Treaty Organization. Thus, in the context of both international economic development and Cold War politics, the signatories were drawn from an extremely homogeneous group. Such a uniform membership obviously facilitated the conclusion of the agreement, but it also served to colour the MTCR as a Western supplier cartel. This has complicated the expansion of the regime.

Three groups of states were excluded from the MTCR: *(1)* major suppliers such as the Soviet Union and the PRC; *(2)* certain European suppliers, some of whom were involved in technology and expertise transfers; and *(3)* second-tier suppliers. The deliberate exclusion of the Soviet Union from the MTCR negotiations was prompted by the poor state of East–West relations, and by fears that Moscow would not permit its missile transfers to be checked and could even act to obstruct the agreement. Yet, clearly, the successful control of international ballistic missile proliferation required the co-operation of the world's largest supplier. Without it, not only could the USSR continue to transfer missiles to the Third World, but the deployment of these weapons by Soviet clients could result in missile arms races, with Western clients coming under increasing pressure to acquire their own systems.

Fears that the exclusion of the Soviet Union from the initial discussions would strengthen Moscow's opposition to the MTCR were vindicated immediately after 1987, when the USSR refused to become a formal party to the regime. In discussions between the superpowers in 1988, Moscow criticized the MTCR for being both too narrow and too broad. It demanded the narrowing of the MTCR technical limits and the development of a broader multilateral regime which would include both suppliers and consumers of missile technologies, and which would link 'states with the advanced technological base of missile production and countries interested in obtaining access to space', such as Moscow's suggested World Space Organization. In return for its membership of the MTCR, the Soviet Union demanded the loosening of the rules of the Coordinating Committee for Multilateral Export Controls so that Moscow could effectively offer its space launch services on the international market, as well as restrictions on the export of aircraft and the inclusion of the PRC in the regime.[29]

Between the period in which the MTCR was announced in April 1987, and the end of 1988, the Soviet Union supplied its accurate SS-21s to both North and South Yemen. Throughout the critical 1988–9 period, during which the Soviets refused to join or formally adhere to MTCR guidelines, hundreds of *Scuds* were transferred to the Afghanistan government – some of which, despite the Soviet withdrawal, are believed to be manned by Soviet troops.[30] These Soviet actions clearly underlined the weakness of the regime. However, as will be discussed later, the USSR has since become more supportive of the MTCR.

The lack of PRC involvement in the MTCR negotiations enabled that country to reach an agreement with Riyadh in the mid-1980s to deliver long-range CSS-2 missiles to Saudi Arabia. There are also continued reports of Chinese involvement in Pakistani, Egyptian and Iranian missile projects.[31] The PRC remains an unpredictable actor in the area of international missile supplies. Although Beijing has provided assurances that it will not transfer any more ballistic missiles to the Third World, many Western analysts have questioned whether the Chinese will, in the long run, demonstrate any great restraint. The need to finance domestic modernization by exporting weapons may encourage the PRC to make new missile transfers. There is no doubt that the failure to secure PRC membership to the MTCR was a major setback for the constraint of ballistic missile proliferation in the Third World in the 1980s.

Many industrialized states remain outside the MTCR. The fact that not all of the European Community (EC) states are members may result in several problems in the areas of export co-ordination and control, especially after 1992, when the Community plans to remove all trade barriers within its borders. Theoretically, since missile technology will be freely transferable within the EC, it will be possible for missile technology originating in an MTCR signatory state to be exported through another EC country. Furthermore, none of the East European states have joined; neither have several West European nations which are military exporters to the Third World. For example, private companies in Austria and Switzerland have become involved in missile technology transactions. Sweden, which has in the past exported large amounts of arms, remains outside the regime.

One of the more formidable obstacles facing the MTCR is the opposition of second-tier suppliers and other Third World missile consumers. Many of these countries regard the regime as discriminatory. The MTCR is viewed as an initiative originating in the Western industrialized world and pursued for the purposes of Western security and scientific interests. Certainly, to the extent that it is an attempt by the North to control the behaviour of the South, it can be argued that the MTCR is essentially discriminatory. US Senator John McCain, a proponent of stronger efforts at constraining proliferation in the Third World, stated that despite such charges, 'US policy must be based on the *de facto* principle that nations do not have the "sovereign right" to proliferate weapons of mass destruction which endanger the broader international community'.[32] The expression of such '*de facto* principles' undoubtedly affects Third World sensibilities.

Consequently, many Third World states have been highly critical of MTCR attempts to control missile proliferation. For example, Calcutta's *Telegraph* was adamant that India would not be 'deflected [from its missile programme] because a few beady eyed gentlemen in Washington have determined the "acceptable" limits of our place in the world'.[33] In an article in the *Times of India* during October 1989, the point was put even more succinctly:

> It is clear . . . that if any one country in the world is compelled to develop missiles because of the proliferation and deployment policies adopted by major missile powers, it is India. It is ironic that having created a major security problem through spreading missiles around this country, India is being indicted by the US for a programme of missile development in response to these threats.[34]

In Argentina, Air Force Chief Marshal of Staff, Lieutenant General Ernesto Crespo, in commenting on the US attitude to Argentina's rocket programme, found it astonishing that Washington 'should believe an attempt by a country it labels as Third World to promote development for the well-being of its inhabitants is a threat to peace'.[35]

It therefore seems unlikely that the more significant Third World producers and second-tier suppliers will readily join the MTCR. The status, security and economic considerations associated with the possession of ballistic missiles make it unlikely that many Third World states will readily abandon their missile projects. Countries such as Brazil and North Korea are unlikely to surrender all missile and missile technology transfer opportunities merely because a number of Western states believe that the spread of ballistic missiles may increase the dangers of nuclear proliferation. Indeed, North Korea has stepped up its efforts to export missile technologies.[36] Given that second-tier suppliers make up an increasing share of the Third World missile technology transfer market, the developing nations' opposition to the MTCR guidelines appears to be one of its more serious drawbacks.

PROBLEMS OF INTERPRETATION

There exists an inherent tension within the MTCR between, on the one hand, the desire to standardize assessments of the nature of dual-capable materials and the intentions of potential recipients; and, on the other, by the fact that the final assessments are left to the supplier.

The MTCR attempts to distinguish between missile technology for use in military programmes and civil space programmes. Similarly, it seeks to hinder the spread of cruise missiles and remotely-piloted vehicles, but much of this technology is used in manned aircraft. The problem is that, in practice, there is no clear distinction between the technologies acceptable to all parties involved and no authoritative body capable of ensuring uniformity of interpretation.

With regard to missile relevant technologies, it should be noted that the MTCR guidelines are very different from the 'trigger list' operating in the field of nuclear non-proliferation. Whereas in the nuclear area there is general consensus on the items and materials needed for 'triggering' a response, there is often no such consensus operating within the MTCR.

France's interest in transferring sophisticated rocket technology to Brazil is a case in point. Despite the strong objections voiced by US policy-makers, the French government insists that it is acting both within the letter and spirit of the MTCR. According to Hubert Curien, French Minister for Research and Technology:

> We will respect the interests of the treaty we have signed but also we have the right to judge on our own to whom we are giving the technology. The US does not deprive itself from doing so . . . What we are telling the American authorities is that the use of technology is for non-military purposes. India and Brazil have the necessary expertise to develop these on their own. It is only a problem of time and money. I do not want any interference in such cases.[36]

Assessing the intentions of recipient states with regard to the employment of dual-use technology is not a problem that the US itself has overcome. In 1989, Washington acknowledged that American companies had transferred technology which could be used for missile production to Iraq in 1985, after it had agreed to adhere to MTCR guidelines. In addition, the proposed US sale of supercomputers to India, Brazil and Israel will, according to one American analyst, accelerate both nuclear and missile proliferation and thereby 'make a mockery of US non proliferation policy'.[37] Supercomputers may speed up both missile and warhead design in countries alleged to be involved in these endeavours.

Not only do dual-purpose technologies pose problems for effective MTCR application, but there are also difficulties associated with the term 'nuclear capable'. The MTCR concentrates on missiles that are deemed 'nuclear capable', but what exactly constitutes a nuclear capable missile remains unclear. It is possible that the 500-kg limit may be too high a cut-off point for a first generation weapon and that Israel, for example, may already be able to manufacture much smaller warheads. Missiles with payloads of less than 500 kg may, in any event, be able to deliver chemical warheads – a capability that could prove highly destabilizing and potentially destructive in many regions. Furthermore, by focusing on 'nuclear capable' missiles, Third World short-range and some Third World medium-range missiles are totally excluded from control. Since the industrialized world is most concerned about the longer-range missiles – that is, the missiles that can attack its territory – this focus underlines the discriminatory nature of the regime.

Within each member country, the implementation of MTCR guidelines involves the co-ordination of various bureaucratic departments, many of which have conflicting objectives. Departments involved in the promotion of exports may find themselves in sharp conflict with those concerned with foreign relations or broad security concerns. Within specific departments, there may be confusion concerning national guidelines, a lack of manpower for enacting the provisions or intra-departmental disputes or bargaining that impinge upon the workings of the regime. Indeed, the uncertainty surrounding the MTCR guidelines is reflected not only between but also within states. The results of these conflicts have sometimes been highly confusing, and have further undermined the credibility of the regime.

For example, a former US Deputy Under Secretary of Defence for Trade and Security Policy has described his attempts to prevent the transfer of a special type of analogue computer to be used for missile testing at Iraq's *Sa'ad* 16 complex. There was a strong conflict of interest between his department and the export-orientated Commerce Department which, for trade reasons, kept information on American sales of military useful equipment secret, and was not prepared to concede the military potential of the consignment.[38] He maintained that the Commerce Department was under pressure from manufacturers to sell the computer; the Department replied that it kept information classified for legal reasons and that it was rigorously enforcing the MTCR.[39] In July 1989, US Senator Jeff Bingaman warned that American agencies were failing to provide adequate supervision when it came to the transfer of technology. He reported that in 1988 the Commerce Department's Office of Export Licenses identified 400 items which met the MTCR's criteria for restriction of export. Yet only 100 of these were referred to the State Department for review.[40] However, these problems pale in significance when compared to those recently encountered in the FRG. The FRG has attempted to expand its industrial base partly through a heavy emphasis on exports. This export orientation has been accompanied, and facilitated by, an unwillingness to place impediments in the way of overseas sales. Insufficient manpower to control exports and a dearth of relevant domestic legislation resulted in the FRG becoming one of the main international centres for private companies wishing to transfer missile, chemical, biological and nuclear technologies.

These difficulties test the efficacy and the credibility of the MTCR. They reflect its failure to identify exact criteria for control and to develop consistent interpretations of end-use objectives. They also point to the possibility that the focus of the regime may be misplaced, in that significant dangers to regional security may derive from the proliferation of shorter-range missiles with smaller payloads.

PROBLEMS OF SECURITY

The MTCR does little to address the security concerns of the developing nations and may in some circumstances even exacerbate them. As noted in the previous Chapter, missiles *per se* are not necessarily destabilizing. Third World states may wish to acquire ballistic missiles not for 'aggressive' purposes but for deterrence. Missiles may be necessary to deter enemy missile attacks or any other form of onslaught. In certain circumstances, deterrence may not be attainable without missiles since aircraft, for example, could have difficulty in penetrating enemy air defence systems.

MTCR range and payload categories penalize certain countries' missile programmes more than others; states which are heavily dependent on external assistance for their projects, or those whose programmes are more vulnerable to external interference will be most affected by the restrictions of the regime. The MTCR could therefore impact on regional power balances in ways that do not necessarily reinforce stability. Arguably, regional crisis stability may be strengthened by proliferation, while the inevitably unequal application of the guidelines may undermine deterrence postures.

Significantly, while the MTCR may hinder the modernization of ballistic missiles in Third World arsenals, it can do little about existing systems, besides making them more difficult and expensive to maintain. Some of these systems could cause great damage if they were ever deployed.

Another problem stems from supplier–client relations. Suppliers' recognition of the security concerns of their clients may lead them to pursue policies which may be interpreted as contravening the letter and spirit of the agreement. The United States, for example, has indicated that it is not prepared to jeopardize its relations with either Saudi Arabia or Israel by threatening to stop arms sales in an attempt to halt their ballistic missile programmes and deployments. In his testimony before a House Sub-committee, Assistant Secretary of State Clarke made this clear when he stated that:

> The real question is: Is that single issue enough to break off a military relationship?. . . I think that if we started this policy [of missile production and military relations linkage] we would lose influence throughout the Middle East very rapidly.[41]

It has been argued that the United States' ambivalent attitude to the Israeli missile project undermines the legitimacy of its missile control efforts in the eyes of other Third World proliferators. The US stands accused of having a double standard: one for its close allies and one for its adversaries.[42] Yet, as Israel's main supporter, Washington cannot and does not, expect Israel to disarm unilaterally when the missile capabilities of its Arab adversaries are increasing and improving.

Similar dilemmas confront the Soviet Union. The need to supply ballistic missiles to clients, or at least to refrain from demanding their removal, may previously have prevented it from joining the MTCR. Yet while the Soviet Union has refused to supply Syria with more capable missiles, it is extremely unlikely that it would expect Damascus to surrender its only guaranteed means of penetrating Israeli air space – even if Israel was prepared to relinquish its systems in a multilateral arms-control agreement. In the final analysis, the MTCR is a set of guidelines for suppliers and is not designed to address the real security concerns of Third World states. In contrast to the NPT, it does not offer Third World countries any rewards for refraining from acquiring missile technologies. The regime thus avoids the pitfalls of the 'dove's dilemma', but does little to attract states whose only means of penetrating enemy air space is through the employment of ballistic missiles, or which require systems to deter or attack distant enemies, or those which believe that a missile capability is needed to deter certain enemy capabilities. In such situations, states will continue to regard the acquisition of ballistic missile systems as essential.

Controlling missile proliferation in the 1990s

ALTERNATIVES TO THE MTCR

The limitations of the MTCR have resulted in a variety of more ambitious arms-control proposals suggested by both academics and policy-makers as alternatives or accompaniments to the existing regime. The essence of many of these proposals is the inclusion of recipient states. This would involve a major diplomatic undertaking, but one which would make much sense, as it is essentially the demand for missiles which is driving countries, companies and individuals into the missile market.

Moscow, for example, has been at the forefront of calls for the establishment of a World Space Organization, a multilateral body under UN auspices which would monitor the development of space vehicles worldwide. This organization would be similar to the IAEA which oversees Third World civil nuclear activities. One of its major objectives would be to ensure that space technologies and programmes were not used for the development of ballistic missiles. It would do so by acting as a channel for space technologies between supplier and recipient, providing launch facilities for states wishing to undertake space research but unable to develop all the means required, and verifying that no transfer of technology from civil to military production occurred in the Third World.

Another approach towards controlling the spread of Third World ballistic missiles would be to universalize certain aspects of the arms-control treaties negotiated between the superpowers. For example, one suggestion is that the INF Treaty, signed in December 1987 between the US and the USSR, should be broadened. Under this proposal, Third World states would pledge not to develop or acquire missiles within a certain range. However, because the range specifications of the INF Treaty are set at 500–5,000 km, it would not include several of the systems currently under development which fall beyond these ranges.

To increase its relevance to the Third World, the parameters of the INF Treaty would have to be extended – possibly in the form of a a ban on NATO and WP short-range nuclear force missile systems. A completely new international regime may have to be established. The scope and content of a Multilateral Ballistic Missile Treaty (MBMT) would be open for negotiation. One analyst has stressed that: 'At the very least it should establish a new standard in international law prohibiting first use of long range ballistic missiles in war'.[43] Ultimately, such a regime could be more ambitious, seeking to 'build-down' existing arsenals and establish confidence-building measures amongst protagonists.

It may not be necessary to develop a complex MBMT in order to curb Third World ballistic missile development. Efforts may be more effectively expended on specific measures to curtail missile development programmes. Military ballistic missile programmes, like nuclear weapon programmes, require extensive testing to ensure reliability of design and confidence that the weapons will work when required. A good control strategy would be to place restrictions on flight testing. As Farooq Hussain has cogently argued:

> The appeal of missile flight-test restrictions as a means of arms control rests largely on the expectation that the innovation of ballistic missiles and the development of new types would then be impossible.[44]

While all these proposals are well intentioned, they still need to take into account that by moving beyond the supply side focus of the MTCR they are bound to encounter vociferous objections from Third World proliferators, most of which would, in any case, ignore such arms-control initiatives. The fact remains that missiles are proliferating in regions where there are substantial tensions and little trust between the protagonists. Fears of enemy intentions, ignorance of enemy capabilities and doubts about the effectiveness of verification procedures make many developing states unlikely candidates for joining multilateral arms-control mechanisms. It was undoubtedly a recognition of such realities that led the United States to concentrate first on the supply side of the market.

A focus on missile recipient states may not only be ineffective but, as noted above, detrimental to the goal of regional stability that must be the central objective of any arms-control regime. By attempting to bring the recipients into the ballistic missile proliferation negotiations, regional tensions will inevitably arise, as numbers are disputed and capabilities debated. It is to be expected that superpower–client relations could come under great strain. Influence needed more urgently elsewhere – for example, in convincing clients to make concessions in the context of broader peace negotiations – could be forfeited. A demand-side focus would also draw attention away from efforts to improve upon the mechanism which is already in place – the MTCR. While the agreement of recipient countries is obviously an attractive goal, two objectives will have to be pursued before it becomes feasible: improved control over the supply of missile technologies and parallel efforts towards improving the regional political climate so as to make regional arms-control a more viable proposition. A wider ballistic missile control regime including recipient states is thus a long-term objective. In the meantime, the inclusion of Third World countries into the ballistic missile arms-control process should take place, where possible, on a regional rather than an international basis and in the context of broader and continuing political and military negotiations.

The MTCR as a 'braking process'

Despite the problems associated with the MTCR, it remains the most appropriate means for restricting proliferation. Although still beset by structural and functional difficulties, it is the most effective system for constraining the spread of missile systems. In assessing the effectiveness of the regime it is important not to impose too high a standard on its functioning and objectives. While it cannot provide an adequate long-term means of controlling missile proliferation, the MTCR is certainly a necessary component of any future strategy. It would be inappropriate to criticize the regime either for its failure to seal all technology transfers, or for 'closing the stable door once the horse has bolted'. The MTCR was conceived in much more limited terms, and in assessing its efficacy, it is necessary to adjust the criteria accordingly.

Essentially, it would be more realistic to view the regime as a 'braking mechanism', enabling a slowing down of the pace of proliferation, so as to mitigate its more dangerous aspects and to allow more stabilizing processes to come in to play. Even if it is accepted that missile proliferation cannot be totally prevented, the need to slow the process down remains vital. Unrestricted access to technology might create major insecurities by giving rise to an extremely rapid increase in the number of systems, as well as improvements in capabilities. By easing the rate of proliferation, the strategic problems associated with it can be made less acute. By raising the political, economic and technological costs of missile acquisition, the MTCR introduces an element of greater control and predictability into the proliferation process.

In the East–West context, the development of nuclear forces took place in an environment which was arguably less tense than many Third World regions, and one in which considerable efforts were directed towards the study of strategic concepts, stabilizing strategies and force postures. One of the more alarming aspects of current missile, chemical and nuclear proliferation is that it generally appears to be isolated from attempts to build upon new concepts of security. During this breathing space, the implications of missile proliferation can be more carefully considered by the regional actors concerned and more stabilizing policies may be pursued by these countries as well as outside powers. By slowing down the proliferation process, the MTCR could grant proliferating states more time in which to grasp fully the strategic implications of the spread of missile systems and perhaps to adopt more stabilizing military stances.

External actors may be able to use the resulting breathing space to help bring about a regional political climate more conducive to arms-control negotiations. The political component is crucial, since a sole or primary focus upon arms-control regimes may only serve to heighten tensions, or divert attention from the basic political problems. As previously noted, the agreement between Brazil and Argentina to allow visits to each other's nuclear facilities, and the accord between Pakistan and India not to attack each other's nuclear sites and to begin discussions on a nuclear test ban, are examples of such confidence-building measures. Should the political and security climate in the regions improve, or should there be negotiations towards resolving regional disputes, various forms of confidence-building measures can be introduced, especially those that help prevent surprise attacks. Confidence-building measures may in turn facilitate political dialogue or military disengagement.[45] Such measures could include: a commitment by the states concerned to notify neighbours of any missile testing; agreements to allow international inspectors to verify that space launch sites are not being used to manufacture long-range missiles; and agreements to share relevant military data.[46]

When the aim of the MTCR is conceived as a 'braking process', the criteria for assessing its success become somewhat less stringent. For example, it is not necessary (although it is, of course, preferable) to prohibit the transfer of all gyroscopes to a specific country in order to restrict radically any missile programme. Large numbers of this item (and others) are needed to test and develop a reliable missile system: a lack of sufficient parts may lead to severe technical difficulties and reduce the confidence of the project leaders. This may lengthen the development process and delay deployment, thereby providing sufficient time in which to introduce security and confidence-building measures and then integrate them with arms-control proposals.[47] Provided that efforts are made to exploit the braking process by engaging the proliferators in confidence-building measures which offer some hope of stabilizing regional political and military relationships, the occasional failure of the MTCR in controlling exports need not be viewed as a total indictment of the regime.

In the final analysis, it should also be remembered that it took four years of secret negotiations, and a great deal of effort and resources to set up the regime. To discard it now in favour of other untested alternatives would be to waste much of this previous investment. Indeed, although the success of the MTCR has so far been limited, a new, more ambitious regime may be more difficult to establish and may ultimately prove to be less effective.

Modifying the MTCR

The necessity of dealing with the MTCR's limitations, the difficulties of presently moving towards a focus on recipients and the wisdom of employing what is already in place, means that further actions to control the proliferation of ballistic missiles should fall within the framework of the MTCR. This appears to be recognized by the policy-makers involved and over the past few years much effort has been directed towards improving the regime.

EXPANDING THE SCOPE

There is no doubt that major efforts will have to be made during the 1990s to increase the membership of the MTCR. At a meeting in London on 5–6 December 1989, an initiative was launched to attract new signatories.[48] During the late 1980s efforts were directed towards increasing the number of EC members and gaining the Soviet Union's adherence to at least the principles of the regime. Since 1987 the MTCR has gained four new members (Spain and the Benelux countries). The momentum appears strong enough to ensure that most of the EC nations will have joined before 1992 – thereby mitigating the problems associated with free internal movement and varying export laws. Industrialized countries outside the EC also need to be targeted for membership. Sweden, which refuses to join the MTCR because of its neutrality, has adopted export laws similar to the MTCR guidelines.[49] Austria, within whose borders companies have engaged in missile technology transfers, should be encouraged to follow the Swedish lead. Furthermore, some way of integrating the countries of Eastern Europe will have to be found. A number of these states, such as Poland and Czechoslovakia, have significant military industries, while Romania is alleged to have been involved in missile projects. In short, the supplier-side focus will have to continue, with particular emphasis on Europe.

The Soviet decision to join would undoubtedly be a major success for the regime and much effort has been given to pursuing this goal. As noted, Moscow first attempted to distance itself from the regime, but now appears to have shifted its position and is, in essence, a *de facto* supporter. At the Wyoming summit in September 1989, the US and Soviet Union 'agreed to activate bilateral consultations on [the missile proliferation] problem'.[50] High-level discussions on this subject have been under way since 1988. Following a meeting between Secretary of State Baker and Foreign Minister Shevardnadze in Moscow in February 1990, a joint statement was issued which included a paragraph stating that they both adhered to the export guidelines of the existing regime with regard to missiles.[51] At the May–June 1990 summit in Washington, the US and the Soviet Union affirmed their support for the objectives of the MTCR; announced that they were taking measures to restrict missile proliferation; were engaging in bilateral consultations to identify measures to prevent missile proliferation; were working to prevent the spread of missiles, especially to the Middle East and would investigate regional initiatives to reduce the dangers of missile proliferation.

Thus, although the Soviet Union has not yet become a *de jure* member, it has clearly signalled its support for the regime. That Moscow is still not prepared to become a full member is to a large degree a reflection of its displeasure at being excluded in the first place. However, in the context of the improved relations between the superpowers, a concerted effort should be made towards overcoming these hesitations and securing Soviet membership. Such an attempt could form part of a broader package of multilateral arms-control initiatives directed at the Third World, covering chemical weapons and conventional arms as well. Soviet membership of the MTCR may facilitate greater control of the international missile market and would do much to boost its image and morale.

While another country to be encouraged to adhere to MTCR guidelines is the PRC, domestic pressures in China to acquire foreign exchange and to modernize the army make the prospects of any firm support unlikely. Rather, the MTCR countries should continue to press Beijing to clarify its missile-transfer policy and to seek firm assurances that it will refrain from transfers to the Third World. In the near term, this may be the only policy option available.

DEALING WITH THE DEMAND SIDE

It remains doubtful whether significant numbers of Third World procurers, second-tier suppliers and other Third World producers will join the MTCR in the next few years. However, this should not prevent the MTCR countries from augmenting the regime by entering into bilateral or multilateral discussions with missile proliferators. As the international scope of the MTCR expands, and its domestic inconsistencies are resolved, its ability to deal with proliferators on a bilateral or multilateral level will be enhanced. Sanctions, *quid pro quos* or regional confidence-building measures will be useful tools with which to constrain demand and mitigate many of the worst effects of missile proliferation.

Quid pro quos may already have been employed in the case of the United States' promise to acquire Argentinian *Pampa* 1A 63s trainer aircraft in exchange for Buenos Aires terminating its work on the *Condor* 2, though this exchange is unconfirmed.[52] Punitive measures (in the form of reductions in aid, military support or declaratory censure) are more costly for all parties concerned, but may become more credible instruments of policy as the MTCR expands. Similarly, multilateral sanctions in the form of economic embargoes on other technology transfers could be effective but may be difficult to apply, even under a larger and more competent MTCR. In establishing priorities for the application of sanctions (or alternatively, for the provision of *quid pro quos*), the MTCR should focus firstly on intra-Third World transfers of missile technologies rather than on domestic production programmes; the former are in many cases motivated more by economic than crucial security concerns and should thus be relatively more susceptible to outside influence.

As noted, the MTCR does not address Third World security issues; regional confidence-building measures, including a focus on ballistic missiles, could help address such concerns. A good precedent for this approach was the US attempt in 1989 to initiate discussions between Israel and Egypt on the subject of ballistic missile proliferation in the Middle East. To reduce fears of a surprise attack, the United States proposed that both sides notify each other in advance of plans for a missile test.[53] While an increase in the number of regional participants seems remote at present, any future attempt to secure a regional settlement will have to include an arms-control component: the Israeli–Egyptian discussions on confidence-building measures could form a basis for any broader approach. These discussions differed from those held between India–Pakistan and Brazil–Argentina on nuclear matters in that they were initiated by an outside power, which happened to be a MTCR member. In a region of potential conflict such as the Middle East, or at times of increased tension, this kind of external intervention may be necessary. As noted, at the May–June 1990 summit in Washington, the superpowers promised to work towards regional initiatives to reduce the threat of missile proliferation.

BUREAUCRACY AND EXPORT LEGISLATION

For the MTCR to function more effectively in the 1990s, greater efforts must be made to tighten domestic export controls. Certain steps are already under way. The streamlining of US controls on the export of missiles is a continuing process. For example, the Pentagon has appointed a Deputy for Non-Proliferation Policy who has consolidated responsibility for missiles as well as nuclear, chemical and biological weapons. Other efforts include: the creation of a new Office of Proliferation in the State Department; the formation of an interagency group to review foreign exports and another to review State and Commerce Department cases; and a strengthening of the Office of Munitions Control.[54]

One signatory to the MTCR which has come under heavy criticism for its lax export laws is the FRG. However, as a result of MTCR pressure, a major review of export policies was begun in 1988. The scope of legislation has been widened, organizations entrusted with monitoring export controls have been strengthened and stricter penalties for violations have been introduced.[55] International arms-control considerations have become a major feature of West German export policy and a firmer stance on ballistic missile technology (as well as nonconventional weapons) can be expected in the 1990s.

More ambitiously, the introduction of more stringent export control procedures will certainly bolster the existing regime. So far, some governments have placed greater emphasis than others on tightening domestic legislation. Inconsistencies between member countries will become more apparent as the international scope of the regime increases. Though a difficult objective to attain, the MTCR should aim therefore at the introduction of a more stringent and common standard of export control.

One of the major challenges confronting the adoption of such a standard is the issue of dual-use technologies. The MTCR has yet to resolve the problem of the relationship between technology transferred for peaceful space use, and that destined for ballistic missile programmes. While a number of ambitious proposals have been outlined for dealing with this issue (such as the broadening of the MTCR to resemble the IAEA or the establishment of a World Space Organization), the present need to avoid entanglement in recipient objections suggests more modest objectives, at least in the short term.

One of these is for transfers to remain the responsibility of the supplier, but as a condition of supply, far more stringent end-use and inspection rights should be applied. If possible, some items to be found in Category 2 should be shifted to Category 1: rocket test facilities and equipment, launch and ground support equipment and certainly some of the more sophisticated types of computers may be suitable candidates. At some stage the broadening of the international scope of the agreement and improvements in domestic legislation will have to be accompanied by a renegotiation of at least some of the MTCR's guidelines, such as the 500-kg payload and the 300-km range critieria.

Of course, renegotiation of the MTCR guidelines may create more problems than it solves and could even exacerbate existing tensions. There is of course no guarantee that a consensus on missile proliferation would emerge, or that the new guidelines would be any more effective. Yet, this remains one area that will need to be more fully investigated if the MTCR is to be strengthened in the next decade.

Conclusion

The recent improvements in superpower relations makes possible the development and strengthening of multilateral arms-control regimes, covering both conventional and non-conventional arms transfers to the Third World. At the Washington summit in May–June 1990, both countries clearly stated that they 'oppose the proliferation of nuclear weapons, chemical weapons, missiles capable of carrying such weapons, and certain other missiles and missile technologies'.[56] It is possible that the 1990s may become the decade in which non-proliferation policy becomes an increasingly important item on the global arms-control agenda.

In this context, international attempts to control the spread of ballistic missile proliferation must be based on the Missile Technology Control Regime. While many attractive alternatives have been put forward, the MTCR remains the most viable base from which to proceed: it is already in place and, while requiring modifications, its assumptions are generally sound. The retention of the MTCR as the centrepiece of international efforts to control the spread of ballistic missiles does not preclude accompanying bilateral or multilateral measures. Indeed, they may work together to raise the costs of missile proliferation. The emerging *détente* between the superpowers may facilitate the development and implementation of such measures.

Built in to the MTCR are a number of values which reflect the belief that missile proliferation is dangerous, that certain Third World activities have to be controlled, and that in the case of missiles, the supplier side of the market must receive primary attention. The resistance of many proliferating states to such a missile technology control regime will only be overcome once they recognize that their security considerations are being taken into account, hence the emphasis on the interaction between arms-control and political processes. However, given some of the mutually exclusive objectives at stake, not all states' interests can be considered equally. This will naturally limit willingness to surrender missile capabilities or interests.

It should also be recognized that superpower readiness to co-operate in the resolution of Third World security problems does not mean that the influence of the US and USSR will increase. As Third World states improve their technological and military capabilities, they may be less willing to tolerate outside interference. Yet, as one analyst has stated, 'The proliferation of ballistic missiles cannot be reversed or eliminated, but some of its more troublesome aspects can be contained'.[57] Modest control objectives certainly need not be militarily or politically inconsequential and – provided they are grounded in a broader political framework – may ultimately help to reinforce regional stability and international security.

CONCLUSIONS

By the year 2000 approximately 20 Third World countries could be deploying ballistic missile systems. States such as India, Brazil and Argentina, which are currently in the development stages of their missile programmes, may have completed production and moved towards deployment by the turn of the century. Other countries that already deploy missile systems may have manufactured new and more capable models. Israel, with its advanced technological base and space programme, could have missiles with accuracies that provide it with a real counterforce capability. In turn, both Israel and India could succeed in deploying missiles with extremely long ranges. Within a decade, many developing countries may possess far more missiles and missile types in their arsenals than at present. The Iraqis and the Brazilians, for example, continue to research and develop missiles with various ranges and payloads. Indeed, if the 1970s and 1980s were characterized by the horizontal proliferation of ballistic missiles in the Third World, the 1990s may see increased vertical proliferation as improvements in missile numbers, range, payload and accuracies steadily take place.

Both status and military rationales will continue to encourage Third World countries to acquire and upgrade their ballistic missile capabilities during the 1990s. According to one analyst, 'As was the case for the West in prior decades, the capability to produce modern weapons has become the *sine qua non* of national sovereignty in the Third World.'[1] Over the past decade, ballistic missiles have become symbolic of Third World technological capabilities and military–industrial expertise. Countries such as Brazil and India have often referred to the prestige associated with their successful rocket launcher programmes. In an international environment in which the influence of the traditional powers appears to be declining, and a number of developing states are searching for new regional and international roles, status symbols such as these may become increasingly important.

The military utility of missiles lies in their ability to penetrate enemy air defences speedily. Third World states may therefore acquire ballistic missiles in order to secure a capability that their air forces could not otherwise provide. For a country such as Syria, whose air force is demonstrably weak, missiles may represent the only option for striking deep inside Israel. Iran, which has an aging air force, may require an extensive missile capability in the next decade. However, the military utility of missiles is limited by the relatively small numbers so far deployed and by system inaccuracies. These limitations continue to make missiles suspect weapons if used against small and hardened enemy military targets. Should accuracies improve and the numbers of missiles in Third World arsenals increase, the military utility of ballistic missile systems may also improve.

During the 1990s, the linkage between missile proliferation and the proliferation of non-conventional ordnance will need to be closely monitored. While it cannot be proved that missile proliferation has accelerated non-conventional proliferation, or *vice versa*, there is clearly a correlation between the two. As missile and non-conventional weapons technologies improve, missiles may be designed to deliver non-conventional ordnance such as chemical, biological and nuclear weapons. Certainly, missiles have proved far more effective when employed not against military targets, but as terror weapons against civilian populations. With the addition of non-conventional warheads, terror weapons may increasingly become weapons of mass destruction. The emergence of a situation in which the ability to impose mass destruction on an enemy will be in the hands of a large number of states – some of which are unstable and unpredictable – will be of immense global significance.

It is uncertain that in an environment of expanding missile and non-conventional arsenals, deterrence will be sustained in a number of Third World regions. While it may be argued that the implications for crisis stability of the proliferation of conventional missiles is not great and that the spread of non-conventional missiles may enhance mutual deterrence, Third World perceptions may be very different to those of Western analysts. Many developing nations face extremely unpredictable strategic environments in which their ignorance of enemy intentions and capabilities is matched only by the certainty of their foes' implacability and, in some cases, willingness to suffer immense casualties. As demonstrated in the 1973 Arab–Israeli War and the Iran–Iraq War, the existence of missiles on both sides did not result in mutual missile deterrence. It is not without significance that both missiles and non-conventional ordnance have been used in the Third World. This suggests that norms on restraint could be overridden in a future crisis. One analyst has maintained that 'interest alone may not halt the escalation from limited to total war . . . [and] [o]nce long-range weapons are in the hands of belligerent states, it may be more difficult to restrain their use through mutual deterrence than a purely rational theory of behaviour would suggest'.[2] Too little is known about the strategic thinking of Third World missile proliferating states to be certain that more missiles or high technology weapons may be stabilizing.

Western security interests cannot remain unaffected by the spread of high technology weapons to the Third World. The ability of an increasing number of developing nations to employ ballistic missiles in order to penetrate Western air space is a new strategic reality that will have to be considered by Western policy-makers in the 1990s. While intentions should not be confused with capabilities, it is not inconceivable that in a crisis situation Third World states could threaten to employ non-conventional missiles against industrialized states. This development will surely affect the relationship between Western and Third World countries and may serve to influence and circumscribe Western strategic options. Even if Third World states use missiles armed with non-conventional warheads against each other, Western interests will be affected, as its allies may suffer serious damage, and norms on employment may be further undermined.

It is for these reasons that the control of ballistic missile proliferation must receive the attention of Western policy-makers during the 1990s. Ironically, at a time when relations between Washington and Moscow have never been better, the ability of the superpowers and the Western industrialized states to control the actions of Third World states is on the decline. Technologically, many developing countries have the capacity to produce vital components of ballistic missile systems either by themselves, or in conjunction with other Third World states, or Western companies and individuals. Consequently, attempts to control the spread of ballistic missile and other conventional and non-conventional technologies can never hope to be entirely successful. Nevertheless, for the foreseeable future Third World countries will continue to be dependent on Western sources for some advanced technologies for their missile projects. By consistently applying the guidelines laid down by the 1987 Missile Technology Control Regime, the industrialized states might be able to slow down the pace of ballistic missile proliferation. This 'braking process' could be vital to the prospects for regional and international security as it would allow a breathing space in which efforts could be directed towards regional military and political confidence-building measures. These might serve to reduce some of the dangers of ballistic missile proliferation and inject a degree of predictability and control into an unstable environment. The key to successful control is an expansion of the MTCR and its more effective application amongst signatory states.

The 1990s will be a period of great change in the international system, as the Soviet Union retrenches and a number of Third World states begin to assert themselves politically and militarily. Should these changes be accompanied by uncontrolled proliferation of highly sophisticated and extremely dangerous weapons systems, the consequences for regional and international security would be grave.

Notes

Introduction

[1] In this study the term 'ballistic missile' is used to describe an unmanned delivery vehicle that may be employed in a surface-to-surface mode, is initially propelled by rocket launchers and follows a parabolic arc for most of the journey towards its designated target.
[2] Speech by W. Webster, 30 March, (London: United States Information Service), 1 April 1989.
[3] Kenneth Waltz, *The Spread of Nuclear Weapons,* Adelphi Paper, no. 171, (London: IISS, 1981).
[4] Hedley Bull, *The Anarchical Society: A Study of Order in World Politics,* (London: Macmillan, 1977), p. 242.

Chapter I

[1] Though now united, the former two states are treated separately in this study.
[2] See for example 'Missile Proliferation in the Third World', *Strategic Survey 1988–1989*, (London: Brassey's for IISS, 1989), pp. 14–24.
[3] There are a number of methods whereby one may attempt to categorize Third World ballistic missiles, for example, according to their payloads or whether they are nuclear capable or not. However, of all the indices available on Third World ballistic missiles, range is the most accessible and clusters of various ranges do occur. (For a similar type of categorization, but employing slightly different range bands, see *ibid.*). It should also be noted that the terms generated in the context of superpower relations, such as ICBM, IRBM etc. are of limited applicability in Third World settings as the distances between protagonists may be much smaller. Thus, when dealing with Third World missiles, the terms short-, medium- and long-range should be employed in a relative sense.
[4] Aaron Karp, 'Ballistic Missile Proliferation', *SIPRI Yearbook 1989: World Armaments and Disarmament,* (Oxford: OUP, SIPRI, 1989), p. 370.
[5] John S. McCain III, 'Proliferation in the 1990s: Implications for US Policy and Force Planning', *Strategic Review,* Summer 1989, pp. 9–20.
[6] Statement by an unnamed official quoted in 'India Proposes Building of ICBM', *Defense and Foreign Affairs Weekly*, 5–11 June 1989, p. 1.
[7] *Foreign Broadcast Information Service* (hereafter *FBIS*) – *AFR*, 'Station commentary', Johannesburg Domestic Radio Service, 0500 GMT, 1 June 1989.
[8] *FBIS–NES*, 'Officials Discuss Missile Capabilities', Teheran Domestic Radio Service, 0745 GMT, 9 March 1988.
[9] 'Israelis Launch Test Satellite With Surveillance Potential', *Washington Times*, 20 September 1988.
[10] *FBIS–NES*, 'Hashemi Rafsanjani Visits Missile Factory', Teheran Television Service, 1705 GMT, 28 March 1988.
[11] *FBIS–NES*, 'Most Iraqi Cities Now Within Missile Range', Teheran IRNA, 29 March 1988.
[12] Roger Fronst, 'Pakistan's New Defence Minister on Missiles, Self Reliance and Afghanistan', *International Defence Review*, April 1989.
[13] 'Saudi Purchase of Chinese Missiles Changes Middle East Military Balance', *Aviation Week & Space Technology* (hereafter *AW&ST*), 28 March 1988.
[14] 'Saudis Warn Iran They May Use Chinese Missiles', *New York Times* (hereafter *NYT*), 29 April 1988.
[15] 'The World's Most Dangerous Man', *US News & World Report*, 4 June 1990.
[16] 'Let Us Not Be Shy Of Agni', *The Hindustan Times*, 19 April 1989.
[17] 'Rabin On Saudi Missiles, Other Issues', *Yedi'ot Aharonot*, 1 April 1988.
[18] Some analysts contend that Israel is developing such systems in order to deter future Soviet power projects in support of Israeli enemies. See Leonard Spector, *The Undeclared Bomb*, (Cambridge MA: Ballinger, 1988).
[19] 'Agni ready for launch', *The Hindu*, 14 July 1988.
[20] 'Saudi Arabia May Sign NPT To Ease Fears Of Its New Missiles', *Arms Control Today*, June 1988.
[21] *FBIS–LAT*, 'Comments From Military Source On Project', Buenos Aires Domestic Radio Service, 1525 GMT, 20 September 1989.
[22] *FBIS–NES*, 'Shomron Discusses Unrest, Saudi Missiles, Syria', Jerusalem

Radio Domestic Service, 0400 GMT, 9 April 1988.

23 Saddam Hussein's reference to 'fire scorch[ing] half of Israel' was accompanied by the statement 'we will hit them with these binary chemical weapons'; see Micha Bar, 'Strategic Lessons of Chemical War', *IDF Journal*, no. 20, Summer 1990, p. 48.

24 A seventeenth could be South Africa which might have received a *Jericho* 1 from Israel.

25 Congressional Research Service Report (hereafter CRS), 'Missile Proliferation: Survey of Emerging Missile Forces', no. 88–642 F, 3 October 1988, p. 64.

26 'Moscow Steps Up Pressure On Syria By Blocking Arms Sales', *The Guardian*, 24 July 1987.

27 'Syrians Defend Missile Purchases', *International Herald Tribune* (hereafter *IHT*), 28 June 1988.

28 'Brazil Offers Libya Missile', *Flight International*, no. 134, 24 September 1989; 'Libya is Developing SSM', *Flight International*, 23 May 1987.

29 'Israel Again asks for Pershing Missile', *NYT*, 7 March 1976.

30 Karp, (*op. cit.* in note 4), p. 389.

31 For a recent discussion of Iraq's missile programmes see 'Iraq's Al Husayn Missile Programme', by J.S. Bermudez Jr. and W. Seth Carus, *Jane's Soviet Intelligence Review*, May 1990.

32 Thomas G. Mahnken and Timothy D. Hoyt, 'Missile Proliferation and American Interests', *SAIS*, Winter–Spring 1990, vol. 10, no. 1, p. 112.

33 For a discussion of North Korean missile development see Bermudez Jr. and Carus, 'The North Korean "Scud B" Programme', *Jane's Soviet Intelligence Review*, April 1989, pp. 177–81. In the mid-1970s North Korea began reverse engineering the *FROG*-7A.

34 For an overview of the *Al-Husayn* see Burmudez Jr. and Carus, (*op. cit.* in note 31).

35 'Fifty Tests For Israel's Rocket', *The Economist*, 14 May 1968, p. 67.

36 CRS, (*op. cit.* in note 25), p. 83.

37 For a discussion of the development of India's TWLRM see Gary Milholin, 'India's Missiles – With a Little Help From Our Friends', *The Bulletin of the Atomic Scientists* (hereafter *BAS*), November 1989.

38 Spector, (*op. cit.* in note 18), p. 38.

39 *Aerospace Daily*, 17 May 1985.

40 'US says Data Suggest Israel Aids South Africa on Missile', *NYT*, 27 October 1989.

41 Karp, (*op. cit.* in note 4), p. 303.

42 For a discussion of the North Korean *Scud* B project see Bermudez Jr. and Carus, (*op. cit.* in note 33).

43 Karl Barth, 'Qadhafi's Secret Missile Oasis: German Technician's Revelations', *Stern*, 21 May 1987, pp. 20–6, 195.

44 'China to update Egypt's missiles', *The Independent*, 14 June 1990.

45 'A case of friendly fallout', *The Economist*, 4 November 1989.

46 The PRC, of course, now manufactures totally domestic systems but has not been considered in this study as one of the newly emerging Third World ballistic missile proliferators.

47 The progress of Libya's domestic missile project is unclear. Syria showed an interest in developing a missile in conjunction with Libya and technicians from the FRG, but since the projects were cancelled in 1981, Syria has not sought to develop its own missile.

48 'Israel's Sputnik', *Christian Science Monitor*, 22 September 1988.

49 G. Steinberg, 'Two Missiles In Every Garage', *BAS*, October 1983.

50 Personal interview.

51 'Officials Comment on Government Nuclear Programmes', *Jornol do Brasil*, 14 December 1983, p. 18.

52 'Indian IRBM Details Revealed', *International Defence Review*, March 1989, p. 247.

53 According to the CRS, 'Thus, the Sonda 1 has a 4.2 kg payload while the SS-07 has one of 4 kg; the Sonda 2 has 44 kg while the SS-40 has a payload of 35–40 kg; the Sonda 3 M1's payload is that of 140 kg and that of the SS-60 140–150 kg'. CRS, (*op. cit.* in note 25), Table 5, p. 92.

54 'Iraq Claims Successful Launching Of Space Rocket', *Jerusalem Post*, 8 December 1989.

55 Radhakrishna Rao, 'Pakistan's Space Ambitions: A Military Option?', *Nature*, vol. 394, 10 December 1981.

56 *Defense Daily*, 7 November 1988.

[57] S. Zaloga, 'Soviet Arms Exports: Cutbacks Ahead?', *Armed Forces Journal International*, December 1989, pp. 20–1.
[58] *Ibid.*
[59] 'CIA Chief Wary Of Pakistan's Nuclear Programme', *NYT*, 19 May 1989.
[60] 'Chinese Sell Saudis Missiles Capable Of Covering Mideast', *Washington Post*, 18 March 1988.
[61] 'Iran: Up In Arms With a Big Hand From The Chinese', *Wall Street Journal*, 5 June 1987.
[62] *The Independent*, (*op. cit.* in note 44).
[63] 'China Denounces Dissenters in Talks with Bush', *Washington Post*, 27 February 1989.
[64] Edward Kolcum, 'Brazil, China Form Space Launch Venture', *AW&ST*, 29 May 1989, p. 35.
[65] See Frank Barnaby, *The Invisible Bomb: The Nuclear Arms Race In The Middle East,* (London: I.B. Tauris & Co, 1989), p. 7 ff.
[66] 'Bush Aides Say Sanctions Could End Israeli Program', *Washington Post*, 31 October 1989.
[67] For further discussion of the transfer of equipment to Iraq, see Alan Friedman, 'The Flight of the Condor', *The Financial Times* (hereafter *TFT*), 21 November 1989.
[68] 'GDR Delegation To Discuss Missile Production', *Al Ittihad*, 5 September 1988, p. 3.
[69] See Andrew L. Ross, 'Do-it-Yourself Weaponry', *BAS*, May 1990, p. 20.
[70] *SIPRI Yearbook 1990: World Armaments and Disarmament*, (Oxford: OUP, SIPRI, 1990), Table 7.1, p. 220.
[71] 'Brazil Industry Depends On Exports', *Washington Post*, 2 November 1986.
[72] 'Avibras Negotiates New Accord With Iraq', *O Globo*, 19 March 1989.
[73] 'Libya Offers To Finance Brazilian Missile Project', *Jane's Defence Weekly*, 6 February 1988.
[74] John Barham, 'Brazil Ignores US Protest Over Arms For Libya', *Sunday Times*, 31 January 1988.
[75] 'US Says Data Suggests Israel Aids South Africa on Missile', *NYT*, 27 October 1989.
[76] 'State Department Sees No Israel–South African Transfer', *Washington Post*, 28 October 1989.
[77] Karp, (*op. cit.* in note 4), p. 295.
[78] 'Iran and Libya in chemical arms for missiles deal', *The Independent*, 30 December 1987.
[79] S. Zaloga, 'Ballistic missiles in the Third World: Scud and beyond', *International Flight Review*, no. 11, 1988, p. 1424.
[80] 'Iran buys more SS 1s', *Flight International*, 25 April 1987.
[81] Karp, (*op. cit.* in note 4), pp. 383, 385.
[82] *Ibid.*, pp. 386, 389.
[83] John K. Cooley, *Libyan Sandstorm*, (Orlando FL: Holt, Reinhart and Winston, 1982), pp. 233–4.
[84] 'Libya Tests Otrag Missile', *Mednews*, 30 November 1987.
[85] For a discussion of the activities of Consen see Friedman, (*op. cit.* in note 67).
[86] 'West Blocks Advance of Condor II Missile', *TFT*, 31 August 1989.
[87] 'Condor II: An Issue To Test US–Egypt Ties', *Middle East Markets*, 17 April 1989.
[88] Friedman, (*op. cit.* in note 67).

Chapter II

[1] Lawrence Freedman, *Arms Control: Management or Reform?,* Chatham House Paper, no. 31, (London: Routledge Kegan Paul; Royal Institute of International Affairs, 1986), p. 6.
[2] 'US Asserts Soviet Advisors are Fighting in Afghanistan', *NYT*, 10 October 1989.
[3] W.S. Carus, 'Missiles in the Middle East: A New Threat to Stability', Research Memorandum no. 6, (Washington DC: The Washington Institute for Near East Policy), June 1989.
[4] 'US calls Libya raid a success', *NYT*, 16 April 1986.
[5] In Europe ballistic missiles did not exist prior to 1939 and the Germans only began employing V-2 rockets in late 1944 – though they may have used them somewhat earlier had it not been for the successful British raid on Peenemünde. For a discussion of the V-campaigns, see David Irving, *The Mare's Nest*, (London: Kimber, 1964).
[6] For a discussion of the Iranian and Iraqi missile campaigns during the Gulf War, see Anthony H. Cordesman and Abraham R. Wagner, *The Lessons of Modern Wars: The Iran–Iraq War*, vol. II, (Boulder CO: Westview Press;

London: Mansell Publishing, 1990), pp. 495–505.
[7] David Ottaway, 'Kabul Forces Gain Combat Edge', *Washington Post*, 27 June 1989.
[8] A. Levran, 'The Growing Threat to Israel's Rear', in *The Middle East Military Balance 1987–1988*, (Boulder CO: Westview Press; Tel Aviv: Jaffee Center for Strategic Studies, 1988), pp. 222–3.
[9] A.H. Cordesman and A. R. Wagner, *The Lessons of Modern Wars: The Arab–Israeli Conflicts, 1973–1989*, vol. 1, (Boulder CO: Westview Press; London: Mansell Publishing, 1990), p. 68.
[10] Fewer than 1,000 V-2s landed in the UK, resulting in approximately 2,800 deaths; 4,000 V-2s landed on Antwerp killing 8,000 people. For a discussion of the V-weapon attacks see Martin Middlebrook, *The Peenmünde Raid*, (London: Allen Lane, 1982). Elsewhere it has been calculated that prior to February 1988 – when Iraq was using original model *Scuds* – each missile succeeded in killing 30–40 Iranians and wounding between 100 and 200; aerial bombing achieved a much higher rate of casualties with some of the more successful attacks during January–February 1987, killing up to 70. See Levran, (*op. cit.* in note 8).
[11] *Ibid.*, pp. 223–4.
[12] A key problem is delivery of sufficient amounts of agent to a large area virtually simultaneously. A co-ordinated missile attack is not easy to organize, especially if launch numbers are relatively few and are dispersed.
[13] It is important to recognize that the number of launchers a state possesses is as significant as the number of missiles it deploys. The more launchers that are deployed, the greater the number of missiles that can be rapidly fired. A lack of launchers may create bottlenecks in the fire plan.
[14] *The Middle East Military Balance 1988–1989*, (Boulder CO: Westview Press; Tel Aviv: Jaffee Center for Strategic Studies, 1989), p. 282.
[15] A. Levran, (*op. cit.* in note 8), p. 223.
[16] *Ibid.*, pp. 223–4.
[17] For a recent outline of the various orders of battle see *The Middle East Military Balance 1988–1989*, (*op. cit.* in note 14), p. 188.
[18] For a discussion of the Israeli and Syrian Air Forces, see Cordesman and Wagner (*op. cit.* in note 9), pp. 243–52; 283–6.
[19] Personal interview.
[20] For a discussion of Israeli active and passive defensive measures, see Cordesman, (*op. cit.* in note 9), pp. 270–3.
[21] M. Navias, 'Ballistic Missile Proliferation in the Middle East', *Survival*, May/June 1989, p. 232.
[22] Personal interview.
[23] Mahnken and Hoyt, (*op. cit.* in Ch 1, note 32).
[24] Carus and Bermudez Jr., 'Iran's Growing Missile Forces', *Jane's Defence Weekly*, 23 July 1988.
[25] 'Soviet Chemical War Chief Alerts West with Syrian Visit', *Washington Times*, 8 April 1988.
[26] W. Seth Carus, 'Chemical Weapons in the Middle East', *Policy Focus*, no. 9, December 1988, pp. 3–4.
[27] Cordesman, (*op. cit.* in note 6), pp. 508–10.
[28] Navias, (*op. cit.* in note 21), p. 233.
[29] Spector, (*op. cit.* in Ch 1, note 18), p. 86.
[30] Ivo Dawnay, 'Brazil and Argentina to Set Seal on Collaboration', *TFT*, 10 December 1986; Richard Kessler, 'Sarney Visit to Pilacaniyeu Was Key to Reciprocal Inspections', *Nucleonics Week*, 23 July 1987, p. 11.
[31] Quoted in David Albright and Tom Zamora, 'India, Pakistan's nuclear weapons: all the pieces in place', *BAS*, June 1989. For a recent scenario of non-conventional war in South Asia see 'India–Pakistan War: It Could Be Nuclear', *NYT*, 7 June 1990.
[32] 'Iraq says it has tested anti-missile missile', *TFT*, 1 December 1988.
[33] D. Gold, 'Star Wars Little Brother', *Jerusalem Post*, 3 December 1988.
[34] Barbara Amouyal, 'DOD Aims at Further Israeli Funding for Arrow Program', *Defence News*, 29 January 1990.
[35] However, given the limited numbers and the time needed for reload of Syrian missiles, the *Arrow* could be quite effective.
[36] *FBIS–WEU*, Coskun Kirca, 'The Question of Syria', *Milliyet*, 27 December 1989, p. 11.
[37] Magnus Clarke, 'Ballistic Missiles in the Third World and the Proliferation of Strategic Defence Technology', *Arms*

Control, vol. 10, no. 2, September 1989, p. 135.
[38] 'Interview with Vice President Dan Quayle', *Armed Forces Journal International*, February 1990.
[39] Doug Brandow, 'The new urgency for SDI', *Washington Times*, 29 March 1990.
[40] James T. Hackett, 'The Ballistic Missile Epidemic', *Global Affairs Magazine*, Winter 1990. The author is a member of the President's General Advisory Committee on Arms Control.
[41] Quoted in R. J. Overy, *The Air War 1939–1945*, (London: Macmillan, 1980), p. 81.

Chapter III

[1] For a general discussion, see John Stanley and Maurice Pearton, *The International Trade in Arms*, (London: Chatto and Windus, for the IISS, 1972). For an early study of the problems of missile proliferation and control see J.H. Hoagland and J. B. Steeple, 'Regional Stability and Weapons Transfer: The Middle Eastern Case', *Orbis*, Autumn 1965, pp. 714–28.
[2] Janne Nolan, 'Ballistic Missile Proliferation in the Third World', *Arms Control Today*, November 1989, p. 12.
[3] The London Guidelines were additional to the so-called 'Zangger Suppliers Guidelines' established under the auspices of the IAEA to interpret Article 111(2) of the NPT. The 'Zangger Guidelines' established criteria for evaluating exports of key components, equipment and materials used in the nuclear fuel cycle which would 'trigger' nuclear safeguards (see IAEA: Vienna, INFCIRC/209). These guidelines did not specifically cover exports of key nuclear technologies like enrichment and reprocessing components, neither did they require IAEA full-scope safeguards as a condition of supply.
[4] The 1976 Symington and the 1977 Glenn Amendments stipulated that the US would not provide military or economic aid to states receiving enrichment or reprocessing technology unless international safeguards were also applied. Both amendments were waived in the 1980s. For a discussion of the 'dove's dilemma' see L.A. Dunn, 'Some reflections on the Dove's Dilemma', *International Organization*, Winter 1981.
[5] Nolan, (*op. cit.* in note 2), p. 11.
[6] Dunn, (*op. cit.* in note 4), p. 185.
[7] See *Nuclear Threat in the Middle East*, (Washington DC: American Enterprise Institute for Public Policy Research, 1975), p. 31.
[8] 'Statement by ACDA Director Iklé to the Conference of the Committee on Disarmament: Arms Transfers, Nuclear Cooperation, and Non-Proliferation, 29 July, 1976', in *Documents on Disarmament 1976*, (Washington DC: US Arms Control and Disarmament Agency, 1978), p. 500.
[9] For a discussion see B. M. Blechman, J.E. Nolan, and A. Platt, 'Negotiated Limitations on Arms Transfers: First Steps toward Crisis Prevention?' in A.L. George (ed), *Managing US–Soviet Rivalry: Problems of Crisis Prevention*, (Boulder CO: Westview Press, 1983), p. 264.
[10] Personal interview.
[11] Personal interview.
[12] 'Missile Technology Control Regime', Statement by the Assistant to the President for Press Relations, 16 April 1987, in CRS Report 1988, p. 104.
[13] Stephen F. von Welck, 'The Export of Space Technology', *Space Policy*, August 1987, p. 230.
[14] Personal interview.
[15] Andrew Gower, Tony Walker and Andrew Macdermott, 'A Deadly Race out of Control', *TFT*, 8 June 1988.
[16] 'Missile Technology Export Control', Statement by the Foreign and Commonwealth Office, 16 April 1987.
[17] Statement by R. Clarke, 'Nations Possessing Ballistic Missiles Could Double', (London: USIS), 2 November 1989.
[18] Quoted in Friedman, (*op. cit.* in Ch I, note 67).
[19] Clarke (*op. cit.* in note 17).
[20] Statement by J. E. Hinds, before the Sub-committee on Defence Industry and Technology of the Commitee on Armed Services, US Senate, 2 May 1989, (Washington: USGPO, 1989).
[21] 'West Blocks Advance of Condor 2', *TFT*, 31 August 1989.
[22] 'US seeks to curb Argentine Missile Program', *Washington Post*, 19 September 1988.
[23] Clark, (*op. cit.* in note 17).

[24] Simon Henderson, 'Iraq Goes it alone on Condor 2', *Mideast Markets*, 15 May 1989.
[25] 'Argentina Abandons Missile Projects', *La Prensa*, 12 May 1990.
[26] Carus and Bermudez, Jr. (*op. cit.* in Ch I, note 31), p. 204.
[27] *Defense and Foreign Affairs Weekly*, 3–9 June 1990; Karp, (*op. cit.* in Ch I, note 4), pp. 377–8.
[28] Spector, (*op. cit.* in Ch I, note 18), pp. 36, 42.
[29] For an outline of the Soviet position see Yuri Pinchikov and Tamara Famusova, 'Prevention of Missile Technology Proliferation', *Disarmament and Security 1988–1989 Yearbook*, (Moscow: IMEMO, Novosti Press Agency, 1989).
[30] *Foreign Report*, 12 April 1990, p. 4.
[31] 'Missile proliferation', *Washington Post*, 30 May 1988; *The Independent*, (*op. cit.* in Ch. I, note 44).
[32] McCain III, (*op. cit.* in Ch I, note 5), pp. 19–20.
[33] Quoted in 'Message Written in Fire', *Time*, 5 June 1989.
[34] 'US Stand on Nuclear Proliferation Scorned', *Times of India*, 24 October 1989.
[35] *FBIS–LAT*, 'Air Force Chief Defends Condor 2', Buenos Aires, 1426 GMT, 21 September 1988; Andrew Slade, 'Condor Project in Disarray', *Jane's Defence Weekly*, 17 February 1990.
[36] 'French Cryogenic Technology for GSLV', *The Hindu*, 30 October 1989.
[37] Gary Milholin, 'Why Are We Helping The Third World Go Nuclear?', *The Washington Post*, 1 April 1990.
[38] Stephen D. Bryan, 'We are the People who let it Happen', *Moment*, December 1989, pp. 53–4.
[39] 'DOD target missile exports', *The Washington Times*, 22 November 1989.
[40] 'Third World Missile Proliferation Emerges as Key Arms Control Issue', *AW&ST*, 3 July 1989.
[41] 'US Aide Opposes Halting Arms Sales Over Missiles', *Washington Post*, 8 November 1989.
[42] See for example, *Report: East–West Co-operation and its Impact on Regional Security Issues*, (New York: Institute for East–West Security Studies, May 1990).
[43] P.G. Schrag, 'A new Genie emerging', *Christian Science Monitor*, 25 April 1988.
[44] Farouq Hussain, *The Future of Arms Control: Part IV. The Impact of Weapons on Test Restrictions*, Adelphi Paper, no. 165, (London: IISS, Spring 1981), p. 19.
[45] See for example Report (*op. cit.* in note 42).
[46] For a discussion of some confidence-building measures, see G. Steinberg, 'The Middle East in the Missile Age', *Issues in Science and Technology*, Summer 1989.
[47] Personal interview.
[48] Foreign and Commonwealth Office Press Release, no. 109, 15 May 1990.
[49] Karp, (*op. cit.* in Ch I, note 4).
[50] 'Documents from the Foreign Ministers meeting', *Arms Control Today*, October 1989, p. 22.
[51] Joint Statement on Non-Proliferation, Washington summit, 31 May–3 June 1990, (London: USIS, 2 June 1990).
[52] 'Condor 2 Cancellation for Pampa Purchase?', *Defence and Foreign Affairs Weekly*, 4–10 December 1989.
[53] *Arms Control Reporter*, 1989, p. 706 B.3-4. and Michael R. Gordon, 'US urges talks on Missiles in Mideast', *New York Times*, 27 December 1988.
[54] Clarke, (*op. cit.* in note 17).
[55] H. Schultze, 'West Germany moves to tighter export controls', *Jane's Defence Weekly*, 1 July 1989.
[56] Joint Statement on Non-Proliferation, (*op. cit.* in note 51).
[57] Nolan (*op. cit.* in note 2), p. 9.

Conclusions

[1] Testimony by Janne Nolan, Hearing before the Sub-committee on Defence Industry and Technology of the Committee on Armed Services, United States Senate, 2 May 1989, (Washington: USGPO), p. 60.
[2] Steven J. Rosen, *The Proliferation of Land-Based Technologies: Implications for Local Military Balance*, ACIS Working Paper, no. 12, (Los Angeles CA: Center for Arms Control and International Security, UCLA, March 1986), pp. 20–1.